Joint Time-Frequency Analysis

Methods and Applications

Shie Qian
Dapang Chen
National Instruments Corporation

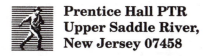
Prentice Hall PTR
Upper Saddle River,
New Jersey 07458

For book and bookstore information

http://www.prenhall.com

Library of Congress Cataloging-in-Publication Data

Qian, Shie, 1949-
    Joint time-frequency analysis: methods and applications / Shie
Qian, Dapang Chen.
        p. cm.
    Includes bibliographical references and index.
    ISBN 0-13-254384-2 (alk. paper)
    1. Signal processing.   2. Time series analysis. 3. Frequency
spectra. I. Chen, Dapang. II. Title.
TK5102.9.Q53 1996
621.382.23--dc20   96-13535
                                                            CIP

Editorial/production supervision: *Diane Heckler Koromhas*
Cover director: *Jerry Votta*
Cover design: *Talar Agasyan*
Manufacturing buyer: *Alexis R. Heydt*
Acquisitions editor: *Bernard Goodwin*
Editorial assistant: *Diane Spina*
Editorial Liaison: *Patti Guerrieri*

 ©1996 by Prentice Hall PTR
Prentice-Hall Inc.
A Simon & Schuster Company
Upper Saddle River, NJ 07458

The publisher offers discounts on this book when ordered in bulk quantities.
For more information, contact;
    Corporate Sales Department
    Prentice Hall PTR
    One Lake Street
    Upper Saddle River, NJ 07458
    Phone: 800-382-3419; Fax: 201-236-7141
    e-mail: corpsales@prenhall.com

The curved line in the "joint time-frequency representation" depicts a radio signal received by the DOE Alexis Satellite. This radio signature is typical of many scientific signals that are hidden by manmade carrier interference and natural noise, which are seen as horizontal lines and periodic marks. JTFA is a vary efficient way to indentify these signals, because neither the "time waveform" nor "power spectrum" reveals the presence of the signal curve seen in the JTFA plot. Screen courtesy of Los Alamos National Laboratory, Group NIS-1.

Printed in the United States of America

10 9 8 7 6 5 4 3 2 1

ISBN 0-13-254384-2

Prentice-Hall International (UK) Limited, *London*
Prentice-Hall of Australia Pty. Limited, *Sydney*
Prentice-Hall of Canada Inc., *Toronto*
Prentice-Hall Hispanoamericana, S.A., *Mexico*
Prentice-Hall of India Private Limited, *New Delhi*
Prentice-Hall of Japan, Inc., *Tokyo*
Simon & Schuster Asia Pte. Ltd., *Singapore*
Editora Prentice-Hall do Brasil, Ltda., *Rio de Janeiro*

To Jun, Nancy, and his parents
-SQ

To His family
-DC

Contents

Chapter 12 Economic Data Analysis   *Ping Chen* *253*

Appendices   *265*

Joint Time-Frequency Analyzer   *281*

Bibliography   *287*

Index   *297*

Preface

The need to analyze a signal in time and frequency domains simultaneously was recognized long ago. But the topic of joint time-frequency analysis (JTFA) other than short-time Fourier transform (STFT) has been largely limited to academic research because of the complexity of the algorithms and the limitations of computing power. Before the '90s, very few modern JTFA methods ran well on the PC.

During ICASSP '91 in Toronto, Canada, advances in the area of JTFA prompted us to develop a software-based joint time-frequency analyzer that would allow one to examine a signal's time and frequency properties together. One year later, in ICASSP '92, San Francisco, National Instruments introduced its first PC-based joint time-frequency analyzer[*]. Besides the STFT-based time-dependent spectrum, the early joint time-frequency analyzer also included several other members of Cohen's class, the Gabor and Wigner-Ville distribution-based time-dependent spectrum (also known as the Gabor spectrogram), and the adaptive representation-based time-dependent spectrum (adaptive spectrogram). Although it was a "Jurassic" version, the first joint time-frequency analyzer attracted a great deal of attention in the industry. Since then, National Instruments' joint time-frequency analyzer has been successfully applied in many areas, such as biomedical engineering, economic data analysis, nondestructive evaluation, radar, RF detection, speech processing, etc. The number of people using this product and the feedback received from users have exceeded our wildest expectation. In retrospect, we completely underestimated the complexity and the level of difficulty in this project. As more and more practicing engineers/scientists explore JTFA, the demand for some comprehensive materials on this subject has become overwhelming. It was this demand that drove us to take on this arduous book project.

Brief Overview of Book

The primary objective of this book is to serve as a professional technical reference and as a senior undergraduate or first-year graduate advanced signal processing textbook. We start each chapter with a discussion of motivation, then theoretical analysis, and end with numerical implementations. All algorithms presented in this book were extensively tested and are suited to student computer projects.

The book contains three parts. In Part 1, we briefly review classical signal analysis that is fundamental for JTFA. The theoretical developments are presented in Part 2. In parallel to the classical Fourier analysis, our presentations are partitioned into linear and bilinear methods. The linear JTFA includes the

---

[*] It was at ICASSP '92 that we, for the first time, presented the *Adaptive Representation*.

Gabor transform (or sampled STFT) in Chapter 3, and the wavelet in Chapter 4. While the Gabor transform and wavelet are evolutions of the linear Fourier transformation, the time-dependent spectrum can be thought of as a counterpart to the classical power spectrum. In Chapters 5 to 8, we discuss several members of Cohen's class, time-frequency distribution series, and adaptive spectrogram. Besides signal analysis, JTFA has also been widely used for detection and estimation of wideband or non-stationary signals. In Chapter 9, we briefly discuss Gabor expansion-based time-variant filtering.

Part 3 contains discussions of three real applications of JTFA, in the areas of radar image processing, biomedical signal processing, and economy and ecology data analysis. All these examples are directly contributed by experts in their fields.

At the end of the book, we have included a demonstration disk for the joint time-frequency analyzer developed by the National Instruments Corporation. The software runs for PC Windows 3.1, Window NT, and Windows 95. The installation procedure is

- Start Microsoft Windows;
- Insert Disk 1 in Drive A;
- From Program Manager, select File menu and choose RUN;
- Type a:\setup and press ENTER.

Unlike the Fourier analysis taught by almost every university, the concepts of JTFA are relatively new and involve some fundamental issues of physics for which no common conclusions have been reached yet. Consequently, the techniques used to achieve JTFA are not standardized. From the application point of view, we have chosen only practically useful techniques. All algorithms introduced in the book were carefully evaluated by authors and National Instruments customers. We omit treatment of some theoretically excellent algorithms that currently are not practical for digital implementations.

Many algorithms introduced in this book rely on advanced levels of mathematics to reach their conclusions. While maintaining the mathematical precision, we have tried to write this book in a style that appeals to the reader's intuition. To achieve this, mathematical rigor and lengthy derivations have been sacrificed in some cases. Hopefully, this style will not unduly offend purists.

We understand that every reader has a different opinion on what he/she reads. Some may find what we have written here to be elementary while others find it overwhelming. Taking this into account, we, as authors, will feel satisfied if our ardor for joint time-frequency analysis rubs off on at least some of those who read this book.

Acknowledgments

Many colleagues and friends deserve thanks for their enthusiasm for this exciting topic as well as their generous help during different stages of the book's preparation. It was their encouragement and help that enabled us to take on such a formidable task. We would like to express our sincere thanks to Ray Almgren, Monnie Anderson, Richard Baraniuk, Michael Cerna, Victor Chen, S.I. Chou, Crystal Doubrava, Mark Dunham, Matthew Freeman, Alexandra Gustafsson, John Hanks, Jonnathan Kim, Shidong Li, Hao Ling, Truong Nguyen, Flemming Pedersen, Qiuhong Yang, and Yunxin Zhao. Our special thanks go to Leon Cohen for serving as one of our most valuable and loyal beta users from the very beginning of the project. His profound questions led us to deep insights into the subject of JTFA.

In a large sense, this book resulted from the cooperative efforts of numerous experts. We deeply appreciate the mathematical proofs, regarding the critical sampling of discrete Gabor expansion and the general convergent properties of adaptive representation, from Zhanbo Yang, Junjing Lei, and E. Ward Cheney. We are enormously grateful to XiangGen Xia. His inestimable help secured a sound mathematical footing for not only Chapter 4, for which he was a primary author, but also many other places.

Likewise, we were very honored by the enthusiastic support and timely responses of radar experts Victor Chen, Hao Ling, Luiz Trintinalia, neurologists Robert Sclabassi, Mingui Sun, and economist Ping Chen in presenting their very interesting applications of JTFA. There is no doubt that their contributions constitute one of the most exciting parts of the book.

Taking this opportunity, we also wish to express our thanks to Prof. Joer J. Morris and Prof. Alen C. Bovik for their support and guidance during our graduate study. It was Prof. Morris who led the first author into this exciting field.

Our thanks should also be expressed to Trish Hill, Michael Hewitt, and Matthew Lesher for their help in teaching us FrameMaker and advanced image processing software. Almost everybody in National Instruments has been very supportive and rendered much-needed assistance.

Finally, this tribute of thanks cannot be concluded without mentioning the support from Dr. Jim Truchard, the president of National Instruments Corporation. Since the beginning of JTFA, Dr. Truchard has been a strong advocate of this project. It is his great enthusiasm and continuous support that permitted us to work in a very pleasant environment and follow through with publication of this book.

钱世锷, 陈大庞

Shie Qian and Dapang Chen
Austin, Texas
January, 1996

Mathematical Conversions

- Continuous and Discrete Variables: We use "( )" and "[ ]" to distinguish continuous and discrete variables. For example, WVD$[k,\theta)$ implies that the variable k is discrete and θ is continuous.
- Dual Functions: In most cases, except for the Gabor expansion, we use "^" to represent the corresponding dual function (e.g., $\hat{\psi}_n(t)$ / $\hat{\psi}_n[k]$ is the dual function of $\psi_n(t)/\psi_n[k]$).
- Periodic Sequence: We use "~" to represent periodic sequence (e.g., $\tilde{f}[n]$).
- Fourier Transform: We use the capital letter functions to denote the Fourier transforms (e.g., $S(\omega)$ and $\tilde{S}[k]$ denote the Fourier transform of $s(t)$ and $s[i]$, respectively).
- Integrals/Summations: All integrals/summations without limits imply integration/summation from minus infinite to plus infinite, i.e.,

$$\int \equiv \int_{-\infty}^{\infty} \qquad \sum \equiv \sum_{-\infty}^{\infty}$$

All multiple integration/summation are assumed exchangeable.
- Matrix Notations: When a capital letter is used to denote a matrix (e.g., A and H), the corresponding lower case letter with subscript ij refers to the (i,j) entry (e.g., $a_{i,j}$ and $h_{i,j}$). We also use the notation $\vec{\gamma}$ to designate a vector, whose ith element is γ_i.

Root

Why Do We Want Joint Time-Frequency Analysis?

From a mathematical point of view, we can describe a given signal in many different ways. For instance, we can write the signal as the function of time, which shows how signal magnitude changes over time. Alternatively, we can also write the signal as the function of frequency by performing a Fourier transformation, which tells us how quickly signal magnitude changed. In real applications, frequency presentations, such as the power spectra, usually have simpler patterns than time waveforms. For example, the complex sinusoidal function corresponds to one pulse in the frequency domain. Therefore, the power spectrum can serve as the "finger print" that proves the existence of some event in which we are interested. To better appreciate the significance of frequency analysis, let's look at a simple example: tuning the piano.

Traditionally, piano technicians use their ears to adjust the difference between the sound generated by a piano and pitch instruments. The calibration results thereby are rather subjective, which largely relies on the technicians' experiences. By the "frequency analysis," however, we could do a much better job.

Fig. 1–1 plots time waveforms of middle C played by a YAMAHA console piano and chromatic pitch instruments. Obviously, in this case, time waveforms do not provide useful information for tuning pianos. From time waveforms, there is no way to tell whether we should tighten or loosen the string. Fig. 1–2 depicts power spectra corresponding to the sound produced by a piano and chromatic pitch instruments. The power spectrum generated by pitch instruments (see Fig.

1–2b) indicates the correct fundamental frequency and harmonics for the standard middle C tone. By comparing it with that produced by a piano (see Fig. 1–2a), the piano technician knows immediately what he should do next. If the fundamental frequency and harmonics are lower than that suggested by pitch instruments, then tighten the string, or *vice versa*. Because spectra describe relative magnitudes of harmonics, using that information, we can further quantitatively evaluate the quality of music instruments. The computerized tuning not only guarantees the quality, but also is less subject to human experiences.

Tuning most musical instruments now could be accomplished by commercial digital tuners that automatically detect the offset between the standard fundamental frequency (the first harmonic) and sound plied by musical instruments. The standard frequency is directly produced by an electronic signal generator inside of the digital tuner rather than the acoustic pitch instruments. Therefore, it has very high accuracy. With the advent of modern personal computers and cheaper electronics, even musical novices can perform the black art of tuning.

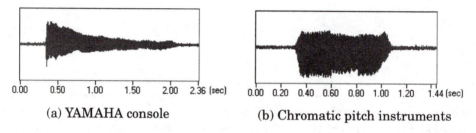

(a) YAMAHA console   (b) Chromatic pitch instruments

Fig. 1–1 Time waveform of middle C tone

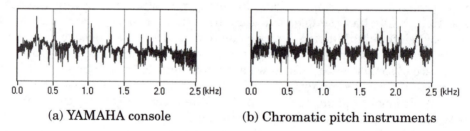

(a) YAMAHA console   (b) Chromatic pitch instruments

Fig. 1–2 Power spectrum of middle C tone

Tuning piano is one of many applications of the signal frequency representation. The reader may name many others. Although frequency representations contain the same amount of information as time waveforms (transformations between the time and the frequency are complete), the frequency representations (e.g., power spectra) are much more useful for piano calibration and many other applications.

A typical goal in signal processing is to find a representation in which certain attributes of the signal are made explicit. In principle, there is an infinite number of ways of describing the given signal. The most important and fundamental variables in signal processing are *time* and *frequency*. A signal's time and frequency representations have been found to be the two most useful signal representations. However, we traditionally have studied signals either as a function of time or as a function of frequency, not both. For instance, we usually run the power spectrum analysis independently of the time factor. This, in spite of the fact that the majority of signals encountered in the real world have time-dependent spectra, such as tones of music that vary with time. In nature, very few signals have frequency contents that do not change over time. In many real applications, it is far more useful to characterize the signal in time and frequency domains simultaneously.

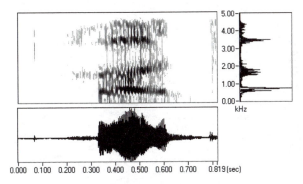

Fig. 1–3 "Hood" (Data courtesy of Y. Zhao, the Beckman Institute at the University of Illinois)

One of the most popular applications of the joint time-frequency analysis (JTFA) may be speech signal processing. Fig. 1–3 shows graphs of a speech signal. The plot at the bottom of the figure is the time waveform of the word "hood" spoken by a five-year-old boy. The plot on the right is the standard power spectrum, which reveals four frequency tones. From the spectrum alone, however, we cannot tell how those frequencies evolve over time. The larger plot in the upper left is the time-dependent spectrum, a function of both time and frequency, which clearly reveals the pattern of the formants. From it, not only can we see how the frequency changed, but we also can see the intensity of the frequencies as shown by the relative brightness levels of the plot. Consequently, by using JTFA, we can better understand the mechanism of human speech.

Recently, it was found that JTFA can also be used to study economical phenomena. Traditionally, economic behavior is considered to be unpredictable. Economic data are commonly treated by techniques developed through probability theory, which essentially declines any possible economic cycles. On the other hand, looking at history, it is not too hard to discover some cycles, though they are changed over time. This observation inspires the researcher to apply JTFA to study economical movements.

Fig. 1–4 depicts the monthly record of S&P 500 from 1947 to 1993. The bottom figure plots the Index against time and indicates that the S&P 500 steadily grew in the last fifty years. We could well approximate the trend by the conventional polynomial curve-fitting algorithm, as shown by the smooth line.

The trend is useful for long-term prediction. But most investors are more interested in short-term behaviors. Therefore, we first compute the fluctuation, the difference between the raw data and the trend, which describes the short-term changes[*]. Then, we compute its standard power spectrum as illustrated in the right plot. Although the standard power spectrum shows that S&P 500 has a four-year cycle (the length of one term of the U.S. presidency), it is not clear whether or not this cycle constantly appears over the entire last half century. From the power spectrum alone, we could not determine if the U.S. economy has been dominated by four-year cycles.

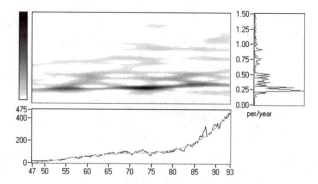

Fig. 1–4 The time-dependent spectrum indicates that the fundamental cycle of S&P 500 is four years, which extends over last the five decades. This reveals the fact that the U.S. economy is strongly influenced by political changes. (Data courtesy of Ilga Prigogine Center, University of Texas at Austin)

Now, let's examine the time-dependent spectrum, the larger plot in the upper left. It indicates that the fundamental cycle is four years, which extends over last the five decades. This reveals the fact that the U.S. economy is strongly influenced by political changes. Economists have for a long time been trying to discover the relationship between the events of the economic system and corresponding outcome. JTFA seems a very promising alternative for analysis methods currently used in the economic society.

In addition to being used for studying the time-dependent spectra, JTFA is also a very powerful tool for removing noise and interference from a signal. In general, random noise tends to spread evenly in the joint time-frequency domain, while the signal itself concentrates in a relatively small range. Consequently, the

---

[*] The key of applying JTFA in economic data analysis is the way of detrending. We shall discuss this issue in more detail in Chapter 12. In this example, the raw data samples of S&P 500 are detrended by the Hodrick-Prescott (HP) filter.

signal-to-noise ratio (SNR) can be expected to be substantially improved in the joint time-frequency domain.

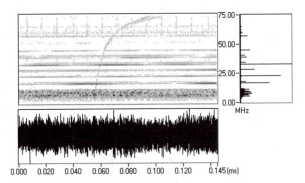

Fig. 1–5 Ionized impulse signal (Data courtesy of Non-Proliferation & International Security Division, Los Alamos National Laboratory)

As an example, let's look at the detection of impulse signals in low orbit satellites. Because the impulse signal may be caused by nuclear weapon testing, the detection and estimation of the impulse signal has been an important national security issue.

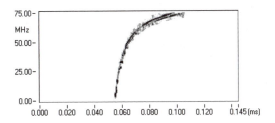

Fig. 1–6 Signal masked from noisy background

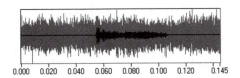

Fig. 1–7 Reconstructed signal

Fig. 1–5 depicts the impulse signal received by the U.S. Department of Energy ALEXIS/BLACKBEARD satellite. After passing through dispersive media, such as the ionosphere, the impulse signal becomes the non-linear chirp signal. While the time waveform is severely corrupted by random noise, the power spectrum is mainly dominated by radio carrier signals that basically are unchanged over time. As shown in Fig. 1–5, neither time waveform nor the power spectrum indicate the existence of the impulse signal. However, when looking at the time-dependent spectrum, we could immediately identify the presence of the chirp-type signal arching across the joint time-frequency domain. By JTFA, recently scientists in Los Alamos National Laboratory have made substantial progress in detecting the radio frequency (RF) non-linear chirp-type signals (see [51] and [153]).

Based on the joint time-frequency representation, we can further mask the desired signal, as shown in Fig. 1–6, then apply the inverse transformation to recover the original time waveform. Fig. 1–7 compares the noisy and reconstructed signals. When the SNR is very low, as with many satellite signals, JTFA may be the only choice to detect and estimate the signal of interest.

The significance of JTFA was recognized as early as the end of World War II. For a long time, JTFA has received attention mainly in academia. Except in the speech processing community, JTFA was not widely accepted. Recently, as the computing power of personal computers rapidly increases, more and more JTFA applications are reported. Unfortunately, for most practical engineers and scientists, JTFA remains a buzz word. It is the aim of this book to systematically discuss the idea and method behind the joint time-frequency analysis and move JTFA into the real world.

This book contains three parts. Because the development of JTFA was closely related to conventional Fourier analysis, the first part of the book briefly reviews classical signal analysis. Although the materials presented in Part 1 may not be new for the reader, the concepts reviewed are very important for our discussions later in the book. We carefully select the topics and examples and use them throughout the remainder of the book.

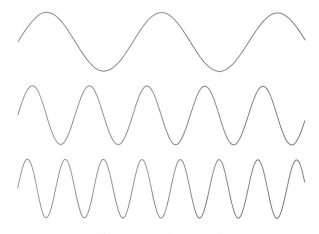

Fig. 1–8 Fourier basis functions

The theoretical developments are presented in Part 2. In parallel to the classical frequency representations, we present the linear as well as bilinear joint time-frequency representations. The linear presentations include the Gabor transform (also known as the short-time Fourier transform) and the wavelets. The fundamental idea behind linear transformations, such as the Fourier transform, Gabor transform, and wavelets, is to compare the signal with a predesigned "ruler." The tick marks of the ruler are made up of the elementary functions. Different elementary functions, or tick marks, lead to a variety of different signal representations.

Fig. 1–8 depicts the tick marks used for the Fourier transformation, which are harmonically related sinusoidal functions. Because each elementary function corresponds to a particular frequency ω, the Fourier transform (the result of comparison between the signal and the tick marks) is the indication of the amount of signal presented at the frequency ω.

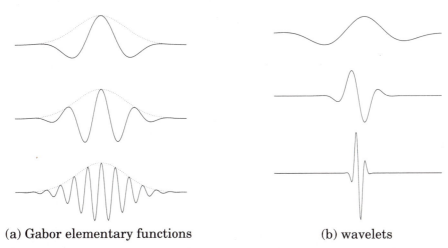

(a) Gabor elementary functions   (b) wavelets

Fig. 1–9 In the Gabor transform, the change of the frequencies is obtained by frequency modulation. In wavelets, the change of the frequency is achieved by scaling time index.

Fig. 1–9 sketches the Gabor and wavelets elementary functions. Unlike the Fourier transform, for which the elementary functions extend into the entire time domain, the Gabor and wavelets elementary functions are centered in a particular time. Therefore, the tick marks employed for the Gabor transform and wavelets not only contain the frequency information but also have time information.

In the Gabor transform, the different frequency tick marks are obtained by frequency modulation. In the wavelet transform, the different frequency tick marks are achieved by time scaling. As shown in Fig. 1–9b, when we suppress the elementary function in time, the oscillating cycles of the elementary functions accordingly decrease, which is equivalent to the increase in the center frequency. In other words, scaling the time indices is equivalent to scaling the frequency in an inverse amount. By systematically changing time scale, we can obtain different frequency tick marks. Because the Gabor elementary functions and wavelets are centered in both time and frequency, the comparison results characterize a signal's behavior in time and frequency simultaneously.

If the Gabor transform and wavelets are evolutions of the classical Fourier transformation, then the time-dependent spectrum can be considered as the counterpart of the power spectrum. The power spectrum is also known as the power density spectrum that describes signal energy distribution in the frequency domain. Unlike the power spectrum, the pattern of the energy distribution in the joint time-frequency domain, such as the Wigner-Ville distribution, is very complicated, which is often associated with high oscillation. Because the high oscillation often obstructs the pattern we expect, traditionally it is considered as interference. One of the central issues in the signal processing community has been how to obtain a time-dependent spectrum that not only possesses a good time-frequency resolution (or well describes a signal's local behaviors) but also has reduced interferences.

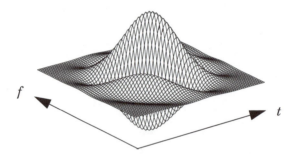

Fig. 1–10 The signal energy in the joint time-frequency domain can be represented in terms of an infinite number of elementary energy atoms. All those atoms are concentrated, symmetrical, and oscillated. The amount of energy contained in each individual atom is inversely proportional to the rate of oscillation.

Recently, it was discovered that by applying the Gabor expansion for the Wigner-Ville distribution, we can break up the signal energy into the infinite number of elementary energy atoms. As illustrated in Fig. 1–10, all those energy atoms are concentrated, symmetrical, and oscillated in the joint time-frequency domain. It is the highly oscillating atoms that bring the interference into the Wigner-Ville distribution. Because the amount of energy contained in each individual atom is inversely proportional to the rate of oscillation, the highly oscillating atoms possess relatively little energy and therefore could be neglected. It has been shown that we can well delineate the time-dependent spectrum by using low oscillating atoms. The resulting representation is named the *time-frequency distribution series* (also known as the *Gabor spectrogram*). All time-dependent spectra illustrated in this chapter are computed by the low order time-frequency distribution series algorithm.

It has been said that *it is with intuition that one invents; it is with logic that one proves.* Just so with the joint time-frequency analysis, the idea seems simple, but the rigorous treatments require a certain level of mathematical preparations. Because this book aims to serve the practicing engineers/scientists and university students who are only expected to understand elementary calculus and linear algebra, we have tried to write this book in a style that appeals to the reader's intuition. We are careful to explain the physical events, while avoiding abstruse mathematics and artificial abstractions. In some cases, mathematical rigor and lengthy derivations have been sacrificed to get to the practical point.

In Part 3, three real applications, in the areas of radar image processing, biomedical signal processing, and economy and ecology data analysis, are discussed in detail. All these examples were directly contributed by experts in their fields. We hope that readers will find reading about these applications both enjoyable and enlightening.

At the end of the book, we include a demonstration disk for the Joint Time-Frequency Analyzer developed by National Instruments Corporation. We are pleased to be able to include the demonstration diskette so that readers can get first-hand experience with the potential of JTFA, and hopefully share some of our enthusiasm!

Review of Signal Analysis

*T*his chapter provides a brief review of the fundamentals to time and frequency analysis, which serves as a quick reference throughout this book. In most cases, we will only present the facts and results without detailed justifications. The reader can find the rigorous treatments in Papoulis [141] and other related references, such as Cohen [32], Oppenheim and Schafer [136], Strang and Nguyen [168], and Vetterli and Kovacevic [180]. Although the materials presented in this chapter may not be completely new, it is certainly beneficial to go through them before reading the rest of book. The concepts and examples introduced in this chapter will be extensively used for the future developments.

In Section 2.1, we briefly discuss the two most important aspects for signal analysis, *expansion* and *inner product*. To study a signal's properties that are not obvious in the time domain, we usually map the given signal from time domain into another domain, in which the interesting properties would be made explicitly. For example, to study the signal's periodic property, we usually write the signal as the sum of harmonically related complex sinusoidal functions $\exp\{j2\pi nt/T\}$. The coefficients (or weights) of each individual elementary function are inner products of the analyzed signal and dual functions. For the Fourier series, the dual functions are the same as the elementary functions $\exp\{j2\pi nt/T\}$, which correspond to impulses at frequencies $2\pi n/T$. Because the inner products reflect the similarity between the signal and the dual functions, Fourier transform indicates the amount of signal presented at frequency $2\pi n/T$.

Obviously, different elementary functions would lead to different signal interpretations. One main topic for signal analysis is how to choose the elementary functions and how to compute the dual functions from the application at hand.

The general expansion theory has a wider scope for deep mathematical
issues, which is beyond the scope of the book. In this book, we only discuss those
expansions that can be realized with the help of a fair amount of linear algebra,
such as the Gabor expansion (or sampled inverse short-time Fourier transform),
orthonormal wavelet, and time-frequency distribution series (also known as
Gabor spectrogram).

In Section 2.2, we review one of the most prominent signal representation
schemes - Fourier transform. The concepts in the classical Fourier analysis are
not only important to analyze the signal whose frequency contents are not
changed in time, but also fundamental to the joint time-frequency analysis.
While the short-time Fourier transform (STFT) is evolved from the traditional
Fourier transform, the time-dependent spectrum, such as STFT spectrogram
and Wigner-Ville distribution, could be considered as counterparts of the conventional power spectrum. Since the Fourier theory has been extensively studied for
many years, in this chapter we will only examine those aspects which are closely
related to the development of joint time-frequency representations.

In Sections 2.3 and 2.4, we review in more detail some important relationships between the time and frequency representations, which are fundamental
for the joint time-frequency analysis. In particular, we discuss the concept of
mean instantaneous frequency and the relationship between frequency bandwidth and time representations. The frequency bandwidth in fact is determined
by variations of mean instantaneous frequency (the derivative of a signal's
phase) as well as magnitude.

Because the time and frequency representations are related via the Fourier
transform, the signal's time and frequency behaviors are not independent. For
example, when a signal's time duration gets narrower, its frequency bandwidth
must become wider. We cannot make the time duration and frequency bandwidth arbitrarily small simultaneously. This assertion is traditionally named
uncertainty principle, which plays an important role in the joint time-frequency
analysis. In Sections 2.5 and 2.6, we give mathematical proof of the uncertainty
principle and the discrete Poisson formula, respectively.

2.1 Signal, Expansion, and Inner Product

The term *signals* generally refers to a function of one or more independent variables, which contain information about the behavior or nature of some phenomenon. The common examples of the signals include electrical current, image,
speech signals, stock indexes, etc., which are all produced by some time-varying
processes. While electrical current, speech signals, and stock indexes are functions of the time, the image signal $s(x,y)$ is a 2D (two-dimensional) light intensity
function, where x and y denote spatial coordinates. The function $s(x,y)$ is proportional to the brightness of image at the point (x,y).

Among the number of infinite possible variables, the most important are *time* and *frequency*, because they are closely related to our everyday life. Based on frequency behaviors, signals can further be grouped into two categories. First is the one whose frequency contents are not changed with time, such as normal engine vibration. It has been well known that the frequency behavior of this kind of signal can be well characterized by the conventional Fourier transform. Very often, people call this type of signal a stationary signal. It is worth noting, however, that the terms *stationary* and *non-stationary* are generally reserved for random signals. Although we shall not delve into the detail of probability theory here, we have to bear in mind, that strictly speaking, it is incorrect to use the word "*stationary*" for the signal whose frequency contents do not change with time.

The second type of signals are those whose frequency contents evolve with time, such as biomedical signals, speech signals, stock indexes, and vibrations. This kind of signal is usually called a *non-stationary signal*. The majority of signals encountered in the real world belong to this category. Because the conventional Fourier transform does not tell how a signal's frequency contents change in time, the classical Fourier analysis is not adequate for many real signals. The goal of this book is to systematically discuss new representations that describe a signal's behavior in time and frequency domains simultaneously.

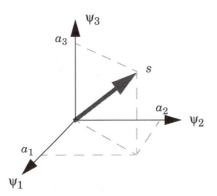

Fig. 2–1 For orthogonal expansion, expansion coefficients a_n are exactly the signal projection on the elementary functions ψ.

From the mathematical point of view, the representations of a signal are not unique. By the *expansion*, we literally can represent a given signal in an infinite number of ways[*]. In other words, given any signal s from the domain Ψ, where Ψ can be finite-dimension or infinite-dimension, we may write signal s in

---

[*] In principle, a given signal can be decomposed in an infinite number of ways. In practice, unless the elementary functions are simple enough, such as orthonormal functions, the realization of desired expansions in general is not trivial. The central topic of this book is the method of extending a time signal as the sum of elementary functions that are concentrated in both time and frequency domains.

terms of linear combination of the set of elementary functions $\{\psi_n\}_{n \in Z}$ for Ψ-domain, i.e.,

$$s = \sum_n a_n \psi_n \tag{2.1}$$

If $\{\psi_n\}_{n \in Z}$ is also in the domain $\hat{\Psi}$, then expansion coefficients a_n in (2.1) depict the signals behavior in the $\hat{\Psi}$ - domain. Fig. 2–1 illustrates an orthogonal expansion, in which the expansion coefficients a_n are exactly the signal projection on the elementary functions.

The most popular example is the Fourier series that decomposes a periodic time signal, $\tilde{s}(t) = \tilde{s}(t + lT)$, where $l = 0, \pm1, \pm2...$, as the linear combination of a set of harmonically related complex sinusoidal functions $\exp\{j2pnt/T\}$. Here, the term *harmonically related complex sinusoidal functions* refers to the sets of periodic sinusoidal functions with fundamental frequencies that are all multiples of a single positive frequency $2\pi/T$, i.e.

$$\tilde{s}(t) = \sum_{n=-\infty}^{\infty} a_n \exp\left\{j\frac{2\pi}{T}nt\right\} \tag{2.2}$$

The formula (2.2) is named the *Fourier series*. The expansion coefficients a_n in (2.2) are a signal's orthonormal projections on the complex sinusoidal functions $\exp\{2\pi n/T\}^*$. They indicate the amount of signals presented at the frequency $2\pi n/T$.

If the set of $\{\psi_n\}_{n \in Z}$ is complete for Ψ, that is, all signals $s \in \Psi$ can be expanded as in (2.1), there will exist a dual set $\{\hat{\psi}_n\}$ such that the expansion coefficients can be computed by the regular inner product, such as

$$a_n \equiv \langle s, \hat{\psi}_n \rangle = \int s(t) \hat{\psi}_n^*(t) dt \tag{2.3}$$

or

$$a_n \equiv \langle s, \hat{\psi}_n \rangle = \sum_k s[k] \hat{\psi}_n^*[k] \tag{2.4}$$

for discrete-time signals. The operations in (2.3) and (2.4) are named *inner products* in the mathematical literature. Formulae (2.3) and (2.4) are also called *transformations*, and $\hat{\psi}_n(t)$ is named the *analysis function*. Accordingly, (2.1) is called *inverse transform* and $\psi_n(t)$ is named the *synthesis function*.

Mathematically, (2.3) and (2.4) are also called inner products and remembered as $\langle s, \hat{\psi}_n \rangle$. The inner product has an explicitly physical interpretation, which reflects similarity between the signal $s(t)$ and the dual function $\hat{\psi}_n(t)$. In

---

* The reader should bear in mind that expansion coefficients a_n in general are not orthogonal projections of the given signal $s(t)$ on elementary functions $\psi_n(t)$ unless $\psi_n(t)$ are equal to its dual function .

other words, the larger the inner product a_n, the closer the signal $s(t)$ to the dual function $\hat{\psi}_n(t)$.

The operation of the inner product in (2.3) and (2.4) may be thought of as using a *ruler*, constituted by a set of functions $\{\hat{\psi}_n\}$, to measure the signal under investigation. Each individual function $\hat{\psi}_n(t)$ can be considered the tick mark of the ruler. The expansion coefficient a_n indicates the weight of the signal's projection on the tick mark determined by $\hat{\psi}_n(t)$. Our everyday experience tells us that the precision of the measurement largely depends on the smallest unit of the instrumentation used. If a physician uses a ruler whose smallest scale is the decimeter to measure a patient's height, then there is no way for the physician to tell the patients height in terms of centimeters. The goodness of the ruler is measured by the fineness of the unit. Therefore, elementary functions should be chosen such that the resulting tick mark is the finest.

If $\{\psi_n\}$ is complete but linearly dependent, the representation is redundant and named the *frame*. In this case, the set of the dual function $\{\hat{\psi}_n\}$ in general is not unique. If $\{\psi_n\}$ is complete and linearly independent, then we say $\{\psi_n\}$ and $\hat{\psi}_n$ are *biorthogonal*. That is,

$$\langle \psi_n, \hat{\psi}_{n'} \rangle = \delta(n - n') \tag{2.5}$$

where

$$\delta[n] = \begin{cases} 1 & n = 0 \\ 0 & \text{otherwise} \end{cases} \tag{2.6}$$

Once $\{\psi_n\}$ is complete and linearly independent, it forms a *basis*.

If $\{\psi_n\}$ is complete and $\langle \psi_n, \psi_{n'} \rangle = \delta[n - n']$, then $\{\psi_n\}$ are *orthonormal*. In this case, $\psi_n = \hat{\psi}_n$. The dual function is itself, which is called *self-dual*. The most well-known example of the orthonormal functions is $\{\exp\{j2\pi nt/T\}\}$. Because the dual functions and the elementary functions have the same form, we can readily obtain the expansion coefficients of the Fourier series in (2.2) by the regular inner product operation (2.3), e.g.,

$$a_n = \int_{-T/2}^{T/2} \tilde{s}(t) \exp\left\{ -j\frac{2\pi}{T} nt \right\} dt \tag{2.7}$$

which indicates the similarity between the signal and a set of harmonically related complex sinusoidal functions $\{\exp\{j2\pi nt/T\}\}$. Because $\exp\{j2\pi nt/T\}$ correspond to impulses in the frequency domain, the ruler used for the Fourier transform possesses the finest frequency tick marks (that is, the finest frequency resolution). The measurements, the Fourier coefficients a_n, precisely describe the signal's behavior at the frequency $2\pi n/T$.

Another well-known orthonormal basis is the sinc function given by

$$\text{sinc}(t) = \frac{\sin(\pi t)}{\pi t} \tag{2.8}$$

which is plotted in Fig. 2–2. It can be shown that $\{\text{sinc}(t\text{-}nT)\}_{n\in Z}$ are complete and orthonormal, i.e.,

$$\langle \text{sinc}(t - nT), \text{sinc}(t - n'T)\rangle = T\delta[n - n'] \tag{2.9}$$

If the signal is band limited, such as $|S(\omega)| = 0$ for $\omega > \pi/\Delta t$, where Δt denotes a sampling interval, then we have

$$s(t) = \sum_{n=-\infty}^{\infty} a_n \text{sinc}(t - nT) \tag{2.10}$$

where

$$a_n = \frac{1}{T}\int s(t)\text{sinc}(t - nT)dt \tag{2.11}$$

which is the sampling of the continuous-time signal $s(t)$ at time nT. Formula (2.10) is usually referred to as the *sampling theory*.

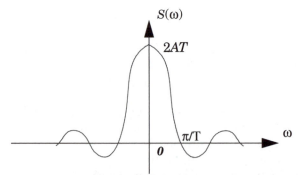

Fig. 2–2 Sinc function

The significance of the orthogonality is twofold. First of all, it has made it easy to compute the expansion coefficients. Referring to (2.3) and (2.4), the computation of the expansion coefficients hinges on the existence of the dual function $\{\hat{\psi}_n\}$. For the set of orthonormal elementary functions, the dual functions are equal to the elementary functions. In this case, the expansion coefficients can be readily obtained, via (2.3) or (2.4), once the elementary functions are determined. However, when the set of elementary functions are not orthonormal, such as $h(t-mT)\exp\{jn\Omega t\}$ (Gabor elementary functions), $\psi((t-b)/a)$ (wavelets), and $\exp\{-\alpha(t-mT)^2-(\omega-n\Omega)^2/\alpha\}\exp\{j(pT\omega+q\Omega t)\}$ (elementary Wigner-Ville distributions), the computation of the dual functions in general is not trivial. Therefore, another central topic of this book is how to implement the expansion when the desired set of elementary functions are not orthogonal.

Secondly, if $\{\psi_n(t)\}$ constitutes an orthonormal basis, then the expansion coefficients a_n are the signal's exact projections on the basis functions $\psi_n(t)$, such as in the case of the Fourier series (2.7). In a biorthogonal case, $\hat{\psi}_n \neq \psi_n$. Because expansion coefficients a_n are inner products of signal $s(t)$ and dual function $\hat{\psi}_n(t)$, they reflect the similarity between the analyzed signal and the dual functions rather than the similarity between the signal $s(t)$ and elementary functions $\psi_n(t)$. If $\hat{\psi}_n(t)$ and $\psi_n(t)$ are significantly different, then a_n may not reflect the signal's behavior regarding the prudently selected elementary functions $\psi_n(t)$ at all. In the joint time-frequency analysis, having the elementary functions $\psi_n(t)$ optimally localized cannot guarantee that the coefficient a_n (transform) also reflects the signal's local behavior. When { $\hat{\psi}_n(t)$} is badly localized in joint time and frequency domain, a_n will not truthfully describe the signal's time-varying nature. We shall see more examples in subsequent chapters.

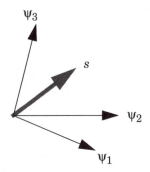

Fig. 2–3 Non-orthogonal expansion

Fig. 2–3 illustrates a non-orthogonal expansion. Although $\{\psi_n\}$ is complete in this case, the expansion coefficients are obviously not equal to the signal projections on $\{\psi_n\}$. When $\{\psi_n\}$ is not orthogonal, we have to first compute the dual function $\{\hat{\psi}_n\}$ and then compute the expansion coefficients a_n via (2.3) or (2.4).

The concepts of the expansion are fundamental for signal representations, which will be used throughout the book. When the expansion is invoked, the first problem is the selection of the desired elementary functions. The word *desired* here implies:

- the elementary functions should have desirable physical interpretations;
- the form of the set of elementary functions $\{\psi_n\}$ should be simple to build.

To discover a signal's property at different frequencies, we hope that the elementary functions are optimally concentrated in the frequency domain, such as $\{\exp\{j2\pi nt/T\}\}$ that correspond to frequency impulses at $2\pi n/T$ and constitute an orthonormal space. In this case, the dual function and elementary function have the same form. The resulting inner product $<s(t),\exp\{j2\pi nt/T\}>$, the Fourier transform, precisely indicates the signal's behavior at the frequency $2\pi n/T$.

To characterize the signal's behavior in the time and frequency domains simultaneously, the elementary functions need to be localized in both time and frequency domains, such as windowed harmonically related complex sinusoidal functions $h(t-mT)\exp\{jn\Omega t\}$ (Gabor elementary functions) or scaled functions

$\psi((t-b)/a)$ (wavelets). Less obvious selection is for the decomposition of the Wigner-Ville distribution.

As introduced in Chapter 5, the Wigner-Ville distribution is traditionally considered a signal energy distribution function in the joint time-frequency domain. It possesses many useful properties for signal processing. The major deficiency of the Wigner-Ville distribution is the so-called cross-term interference. It is observed, however, that the cross-terms are generally localized and highly oscillated. On the other hand, the useful properties of the Wigner-Ville distribution are all obtained by *averaging* the Wigner-Ville distribution. These observations suggest that if the Wigner-Ville distribution is thought as the sum of localized 2D (time and frequency) harmonic functions, then useful properties of the Wigner-Ville distribution will mainly depend on the low oscillated harmonics. This is because low harmonics have larger averages. The high harmonics not only have limited influence on the useful properties, but also directly relate to the cross-term interference. Therefore, they are less important to the joint time-frequency representations. To retain the useful properties and suppress the cross-terms, we should decompose the Wigner-Ville distribution as the linear combination of localized 2D harmonic functions, such as $\exp\{-\alpha(t-mT)^2-(\omega-n\Omega)^2/\alpha\}\exp\{j(pT\omega+q\Omega t)\}$ (2D Gaussian function). The resulting expansion is named *time-frequency distribution series* or *Gabor spectrogram*.

Moreover, it is highly desired that the ruler be easier to build. In addition to the Fourier series, good examples include the Gabor expansion, wavelets, and the time-frequency distribution series. For the Gabor expansion, the set of elementary functions is constituted by time-shift and frequency-modulated single prototype functions $h(t-mT)\exp\{jn\Omega t\}$. In the wavelet presentation, different time and frequency tick marks are accomplished by translating and dilating a single mother wavelet $\psi((t-b)/a)$. In all those cases, once we decide the mother function, the entire set of elementary functions can be readily obtained.

Once the elementary functions are selected, the remaining question is how to compute dual functions. In most applications, desirable elementary functions are neither orthonormal nor biorthogonal. Consequently, implementations of desired expansions are not as simple as the Fourier series. In those cases, the dual functions are not unique. To obtain a best measurement, we have to impose certain constraints on dual functions. One main topic of this book is how to compute the dual function $\{\hat{\psi}_n\}$ for the given meaningful elementary function $\{\psi_n\}$.

The general expansion has a wider scope for deep mathematical issues. For example, the proof of the completeness in general is not straightforward. In this book, we shall limit our discussions to the cases where expansions can be realized with the help of a fair amount of elementary linear algebra.

Finally, it is worthwhile to note that the dual functions ψ_n and $\{\hat{\psi}_n\}$ are exchangeable, that is,

$$s(t) = \sum_n \langle s, \hat{\psi}_n \rangle \psi_n(t) = \sum_n \langle s, \psi_n \rangle \hat{\psi}_n(t) \tag{2.12}$$

Which one, $\psi_n(t)$ or $\{\hat{\psi}_n(t)\}$, is used for the analysis function to compute the expansion coefficients depends on the applications at hand. If we mainly are interested in the expansion coefficients, such as most applications of the short-time Fourier transform and wavelet transform, then we may use $\psi_n(t)$ for the analysis function, because it is selected first and thereby easier to make it meet our requirements. In this case, once $\psi_n(t)$ is properly selected, the expansion coefficients a_n will guarantee to give good estimations of signal behaviors in the joint time-frequency domain.

2.2 Fourier Transform

Although a given signal can be represented in many different ways, the most important are the *time* and *frequency* representations. The significance of the quantity *time* is easy to understand, because it is fundamental. The majority of signals encountered in our everyday life are directly related to *time*. The frequency representations, on the other hand, were not popular until the early 19th century when Fourier first proposed the harmonic trigonometric series. Since then, the frequency representation has become one of the most powerful and standard tools for studying signals. By using frequency representations, we could better understand many physical phenomenon and accomplish many things that cannot be achieved based on time representations. For example, based on spectra, we could better tune and evaluate musical instruments as introduced in the beginning of Chapter 1. By examining the spectrum, we could also identify elements contained in the materials that we are interested in. This is because all elements have their own distinct frequencies. Iron has a different frequency spectrum than that of copper.

By the expansion theorem, the frequency representations of non-periodic signal $s(t)$ is

$$s(t) = \frac{1}{2\pi}\int S(\omega)\exp\{jt\omega\}d\omega \qquad (2.13)$$

where

$$S(\omega) = \int s(t)\exp\{-j\omega t\}dt \qquad (2.14)$$

Eq.(2.14) is named *continuous-time Fourier transform*. $S(\omega)$ is the measure of the similarity between the signal $s(t)$ and complex sinusoidal functions as shown in Fig. 2–4.

One of the most important features of the Fourier transform is that the basis functions in (2.13) and the dual functions in (2.14) have the same form $\exp\{j\omega t\}$. Moreover, $\exp\{j\omega t\}$ corresponds to an impulse at frequency ω. Hence, the ruler used to measure a signal's frequency properties possesses the finest tick marks. Consequently, $S(\omega)$, the signal's projection on the basis functions, precisely reflects the signal's behavior at frequency ω.

The square of the Fourier transform $|S(\omega)|^2$ is called *power spectrum*, which indicates how the signal energy is distributed in the frequency domain.

While the Fourier transform $S(\omega)$ is a linear function of the analyzed signal, the power spectrum $|S(\omega)|^2$ is quadratic to the signal $s(t)$. The Fourier transform $S(\omega)$ in general is complex, whereas the power spectrum $|S(\omega)|^2$ is always real. The Fourier transform and the power spectrum are the two most important tools for frequency analysis.

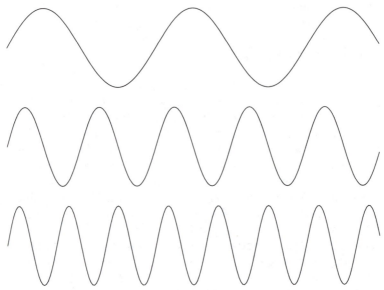

Fig. 2–4 Real parts of complex sinusoidal functions

According to the Wiener-Khinchin theorem, the power spectrum can also be written as the Fourier transform of the signal's auto-correlation function, i.e.,

$$|S(\omega)|^2 = \int R(\tau)e^{-j\omega\tau}d\tau \tag{2.15}$$

where the auto-correlation function $R(\tau)$ is computed by

$$R(\tau) = \int s(t)s^*(t-\tau)dt \tag{2.16}$$

The presentation (2.15) is very useful, which leads to a feasible way of designing the joint time-frequency representations. For example, if we make $R(\tau)$ time-dependent, such as $R(\tau,t)$, then the resulting Fourier transform manifestly is the function of time and frequency, i.e.,

$$P(t, \omega) = \int R(t, \tau)e^{-j\omega\tau}d\tau \tag{2.17}$$

which links the power spectrum to time. Hence, we name $P(t,\omega)$ the time-dependent spectrum. Good examples include the STFT spectrogram as well as the Wigner-Ville distribution. We shall elaborate on this subject in more detail in Chapter 5.

For applications in digital signal processing, it is necessary to extend the continuous-time Fourier transform to discrete-time signal. Let $s[k\Delta t] = s(t)$, where Δt denotes the sampling interval. Without loss of the generality, let $\Delta t = 1$. Then, the discrete-time Fourier transform is defined as

$$\tilde{S}(\theta) = \sum_k s[k]\exp\{-j\theta k\} \tag{2.18}$$

where $\theta = \omega\Delta t$. $\theta/2\pi$ is named the normalized frequency. The inverse discrete-time Fourier transform is

$$s[k] = \frac{1}{2\pi}\int_{-\pi}^{\pi} \tilde{S}(\theta)\exp\{jk\theta\}d\theta \tag{2.19}$$

Because the time variable of $s(t)$ is digitized, its frequency counterpart becomes the periodic function at the frequency domain, that is, $\tilde{S}(\theta) = \tilde{S}(\theta + 2\pi l)$, for $l = 0, \pm1, \pm2....$ Note that the frequency variable θ in (2.18) and (2.19) is a continuous variable.

When the digital computers are used, $\tilde{S}(\theta)$ can only be evaluated at discrete points, such as $\theta = 2\pi n/L$ where $0 \leq n < L$. In this case, (2.18) and (2.19) have to be further modified. The resulting transform is named the *discrete Fourier transform* (DFT).

$$\tilde{S}[n] = \sum_{k=0}^{L-1} \tilde{s}[k]W_L^{-nk} \tag{2.20}$$

where

$$W_L = \exp\left\{j\frac{2\pi}{L}\right\}$$

Because we sample the frequency variables of θ, the time samples in (2.20) further reduce to periodic, that is, for $l = 0, \pm1, \pm2....$ The inverse DFT is

$$\tilde{s}[k] = \frac{1}{L}\sum_{n=0}^{L-1} \tilde{S}[n]W_L^{nk} \tag{2.21}$$

Example 2–1   Complex sinusoidal function

If

$$s(t) = \exp\{j\omega_0 t\}$$

then

$$S(\omega) = \int\exp\{j\omega_0 t\}\exp\{-j\omega t\}dt = 2\pi\delta(\omega - \omega_0) \tag{2.22}$$

For the discrete-time signal,

$$s[k] = \exp\left\{j\frac{2\pi n'}{L}k\right\} = W_L^{n'k} \tag{2.23}$$

the corresponding DFT is

$$\tilde{S}[n] = \sum_{k=0}^{L-1} W_L^{n'k} W_L^{-nk} = \frac{1}{L}\delta(n - n' - lL) \tag{2.24}$$

$l = 0, \pm1, \pm2....$

In the frequency domain, the sinusoidal-type functions are perfectly localized, since their spectrum loads only two points. In the time domain, these function are not localized. Therefore, they are not suitable to analyze or synthesize complex signals presenting fast local variations such as transients or abrupt changes.

Example 2–2   Real cosine function

If
$$s(t) = \cos\{\omega_0 t\}$$

then,

$$S(\omega) = \int \cos(\omega_0 t)\exp\{-j\omega t\}dt$$

$$= \frac{1}{2}\int [\exp\{j\omega_0 t\} + \exp\{-j\omega_0 t\}]\exp\{-j\omega t\}dt$$

$$= \pi\delta(\omega - \omega_0) + \pi\delta(\omega + \omega_0)$$

Unlike its complex counterpart in Example 2–1, the real-valued cosine function corresponds to two pulses at frequencies $-\omega_0$ and ω_0. It is easy to verify that for an arbitrary real signal $s(t)$, its Fourier transform must satisfy $S(\omega) = S^*(-\omega)$. In other words, its power spectrum is symmetry with respect to $\omega = 0$.

Example 2–3   Rectangular pulse signal

If

$$s(t) = \begin{cases} A & |t| < T \\ 0 & |t| > T \end{cases}$$

as shown in Fig. 2–5.

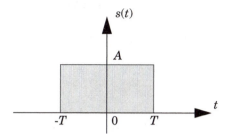

Fig. 2–5 Rectangular pulse

Then,

$$S(\omega) = A\int_{-T}^{T} \exp\{-j\omega t\}dt = -\frac{A}{j\omega}\exp\{-j\omega t\}\Big|_{-T}^{T} = 2A\frac{\sin\omega T}{\omega} \tag{2.25}$$

which is plotted in Fig. 2–6.

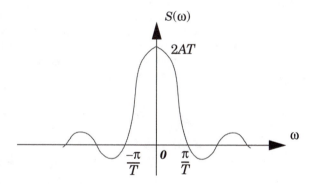

Fig. 2–6 Fourier transform of rectangular pulse

In applications, we usually write the Fourier transform of the rectangular pulse in terms of the sinc function in (2.8). Then,

$$S(\omega) = 2A\frac{\sin\omega T}{\omega} = 2AT\frac{\sin\omega T}{\omega T} = 2AT\operatorname{sinc}\left(\frac{\omega T}{\pi}\right) \tag{2.26}$$

The rate of oscillation of $S(\omega)$ in frequency domain is inversely proportional to the width of rectangular pulse T. The narrower the width of $s(t)$ is, the higher the oscillation of $S(\omega)$.

In contrast to the sinusoidal functions, the rectangular pulse is finitely supported in time domain but badly localized in the frequency domain.

Example 2–4   Gaussian function

$$s(t) = \sqrt{\frac{\alpha}{2\pi}} \exp\left\{-\frac{\alpha}{2}(t-t_\mu)^2\right\} \tag{2.27}$$

which is the standard Gaussian function with the mean time t_μ. The Fourier transform is

$$S(\omega) = \int \sqrt{\frac{\alpha}{2\pi}} \exp\left\{-\frac{\alpha}{2}(t-t_\mu)^2\right\} \exp\{-j\omega t\} dt = \exp\left\{-\frac{1}{2\alpha}\omega^2 + j\omega t_\mu\right\} \tag{2.28}$$

The formula (2.28) usually is remembered as the *Gaussian characteristic function*. Eq.(2.28) shows that the Fourier transform of the Gaussian function is also Gaussian. The time-shift in (2.27) corresponds to the phase-shift in (2.28). Moreover, the variance of the frequency representation in (2.28) is the reciprocal of the time variance in (2.27). The narrower the spread in the time domain is, the wider the spread in the frequency domain, or vice versa. Unlike the sinusoidal or rectangular pulse functions discussed earlier, the Gaussian function is localized in both time and frequency. As we shall see later, among all possible functions, the Gaussian function is optimally concentrated in joint time and frequency domain.

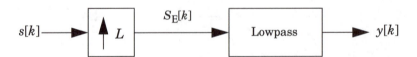

Fig. 2–7 Interpolation Filter

Example 2–5   Interpolation filter

Fig. 2–7 is a block diagram of an interpolation filter. The left block is commonly called an L-fold expander. The right block is a conventional lowpass filter. The expander inserts L-1 zeros between each sample $s[k]$ and produces the output $s_E[k]$ as

$$s_E[k] = \begin{cases} s[k/L] & k = 0, \pm 1, \pm 2 \\ 0 & \text{otherwise} \end{cases}$$

Then

$$\tilde{S}_E(\theta) = \sum_k s_E[k]\exp\{-j\theta k\} = \sum_k s_E[kL]\exp\{-jL\theta k\}$$

$$= \sum_k s[k]\exp\{-jL\theta k\} = \tilde{S}(L\theta)$$

which implies that $\tilde{S}_E(\theta)$ is an L-fold compressed version of $\tilde{S}(\theta)$ as demonstrated in Fig. 2–8. The multiple copies of the compressed spectrum are usually called *images*. If applying a lowpass filter after the expander, as show in Fig. 2–8, then we obtain upsampled signals.

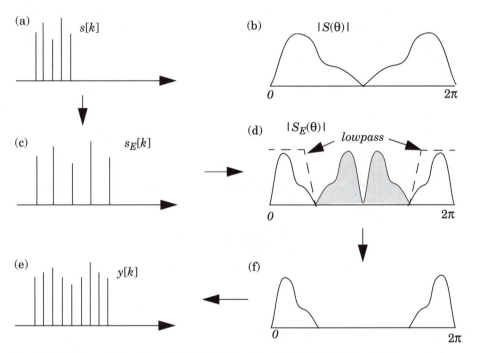

Fig. 2–8 Interpolation (a) $s[k]$, (b) $|S(\theta)|$, (c) $s_E[k]$, (d) $|S_E(\theta)|$, (e) Upsampled samples $y[k]$, (f) Image removed spectrum.

The operation of the interpolation filtering can be described by

$$y[k] = \sum_{n=0}^{N} s[n]\gamma[k-nL] \tag{2.29}$$

where $\gamma[k]$ denotes the lowpass filter with the cut-off frequency at π.

The interpolation filter is a very important filter, which can be used to change sampling rates of the existing digital samples. As introduced in Chapter

5, the sampling rate of the discrete Wigner-Ville distribution usually requires four times faster than the signal's bandwidth, which is twice the conventional sampling frequency. Applying the interpolation filter discussed in this example, we can resolve aliasing without altering the existing hardware structure.

The deficiency of the classical Fourier analysis is that the complex sinusoidal basis functions are not concentrated in time domain (see Example 2–1). The Fourier transform, inner product between the analyzed signal and complex sinusoidal functions, does not explicitly associate with time. Based on the Fourier transform alone, it is not clear whether or not the signal's frequency contents are changed in time, though the phase of $S(\omega)$ is related to time-shift. On the other hand, the frequency contents of the majority of signals encountered in our everyday life are time-dependent, such as the economic index, medical data, the monthly rainfall record, sunlight, and speech signals. It has been recognized for a long time that the time or frequency representations alone are not adequate for many applications. It is beneficial to study the signals in time and frequency domains simultaneously.

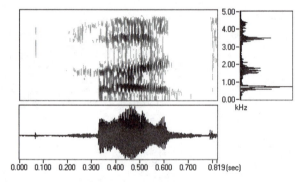

Fig. 2–9 "Hood," spoken by a five-year-old boy (Data courtesy of Y. Zhao, the Beckman Institute at the University of Illinois). The power spectrum (right) indicates there are four frequency clusters, but it does not tell how they are changed over time. In contrast, the time-dependent spectrum (top left) clearly depicts how those four formants vary with time.

Fig. 2–9 illustrates the speech analysis results of an utterance by a five-year-old boy. The power spectrum (right) indicates there are four frequency clusters, but it does not tell how they are changed over time. In contrast, the time-dependent spectrum (top left) clearly depicts how those four formants vary with time. Therefore, the time-dependent spectrum is more valuable for the study of speech pathology than the power spectrum.

It is the goal of this book to introduce joint time-frequency representations that characterize signals in time and frequency simultaneously. Before discussing joint time-frequency analysis, we shall further review the relationship between a signal's time and frequency representations.

2.3 Relation of Time and Frequency Representations

In the preceding section, we briefly reviewed the concepts of the expansion for converting a given signal from one domain to another domain. In particular, we discussed the Fourier transform. Because *time* and *frequency* play the prominent roles in our everyday life, it is important and beneficial to further investigate the relationship between the time and frequency representations. Since the subject of the Fourier analysis has been exhaustively studied for many years, in this section we only exam those concepts that are directly related to our future developments. Moreover, most of the time, only the continuous-time cases are discussed. We leave their discrete counterparts for the reader to exercise.

Shifting   Shift in time by t_0 results in multiplication by a phase factor in the frequency domain, i.e.,

$$s(t - t_0) \leftrightarrow \exp\{-j\omega t_0\}S(\omega) \tag{2.30}$$

Conversely, a shift in frequency by ω_0 results in modulation by a complex exponential in the time domain, i.e.,

$$S(\omega - \omega_0) \leftrightarrow \exp\{j\omega_0 t\}s(t) \tag{2.31}$$

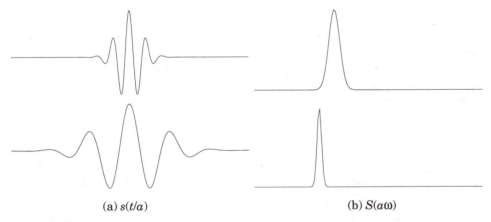

(a) $s(t/a)$ (b) $S(a\omega)$

Fig. 2–10 The top plots correspond to the small scaling factor a. The bottom plots correspond to the large scaling factor a. Scaling in time domain leads to the inverse scaling in frequency domain.

Scaling   Scaling in the time domain leads to inverse scaling in frequency domain, i.e.,

$$s(t/a) = S(a\omega) \tag{2.32}$$

where a is a real constant. Fig. 2–10 illustrates $s(t/a)$ and corresponding Fourier transform. It is interesting to note that the signal's center frequency increases when we suppress the signal in the time domain, or vice versa. In other words, we can systematically adjust the signal's frequency centers simply by scaling the time variables. This suggests that instead of using the complex sinusoidal functions, such as in the Fourier transform, we may employ the scaled functions to build tick marks to measure a signal's frequency behavior. The resulting representation offers many interesting properties and named wavelets. We shall discuss it in great detail in Chapter 4.

Eq.(2.32) also implies that the Fourier transform pair cannot both be of short duration. When the time duration gets larger, the frequency bandwidth must be smaller or vice versa. This assertion, known as the *uncertainty principle*, can be given various interpretations, depending on the meaning of the term "duration." Because the uncertainty principle plays an extremely important role in the joint time-frequency analysis, we shall treat it separately in Section 2.5.

Conjugate Function

$$s^*(t) \leftrightarrow S^*(-\omega) \tag{2.33}$$

If $s(t)$ is real, then $s(t) = s^*(t)$. Hence, $S(\omega) = S^*(-\omega)$, which is named *Hermitian function*.

Derivatives

$$\frac{d^n}{dt^n}s(t) \leftrightarrow (j\omega)^n S(\omega) \tag{2.34}$$

This is a very useful relation, which will be extensively used in this book to convert the function from time domain to frequency domain, or *vice versa*.

Convolution Theorem   For signals $s(t)$ and $\gamma(t)$, the convolution is defined as

$$s(t) \otimes \gamma(t) = \int s(\tau)\gamma(t-\tau)d\tau \tag{2.35}$$

If the Fourier transform of $s(t)$ and $\gamma(t)$ are $S(\omega)$ and $G(\omega)$, respectively, then the convolution of $s(t)$ and $\gamma(t)$ is equal to the inverse Fourier transform of $S(\omega)G(\omega)$, that is,

$$\int s(\tau)\gamma(t-\tau)d\tau = \frac{1}{2\pi}\int S(\omega)G(\omega)e^{j\omega t}d\omega \tag{2.36}$$

Conversely,

$$\frac{1}{2\pi}\int S(\Omega)G(\Omega-\omega)d\Omega = \int s(t)\gamma(t)e^{-j\omega t}dt \tag{2.37}$$

Parseval's Formula   Setting $t = 0$ in (2.36), we obtain

$$\int s(\tau)\gamma(-\tau)d\tau = \frac{1}{2\pi}\int S(\omega)G(\omega)d\omega \tag{2.38}$$

With $\gamma(-t) = h^*(t)$, we have $G(\omega) = H^*(\omega)$. Then, (2.38) reduces to

$$\int s(\tau)h^*(\tau)d\tau = \frac{1}{2\pi}\int S(\omega)H^*(\omega)d\omega \tag{2.39}$$

which is known as *Parseval's formula*. If $s(t) = h(t)$, we have $S(\omega) = H(\omega)$, and (2.39) becomes

$$\int |s(t)|^2 dt = \frac{1}{2\pi}\int |S(\omega)|^2 d\omega \tag{2.40}$$

which implies that the Fourier transformation is energy conserved.

2.4   Time Duration and Frequency Bandwidth

In the preceding section, we reviewed some basic connections between a signal's time and frequency representations. The most important relationship in terms of joint time-frequency analysis, however, is the relationship between a signal's time duration and frequency bandwidth. The concepts introduced in this section play significant roles in joint time-frequency analysis.

As we know, the time duration and frequency bandwidth can be defined in several different ways. The different definitions will lead to the different interpretations. In this book, we employ the most common definition - the standard deviation, the concept used in the probability theorem to characterize the time duration and frequency bandwidth.

Let's define that the energy contained in signal $s(t)$ is E, that is;

$$\|s(t)\|^2 = \int |s(t)|^2 dt = \frac{1}{2\pi}\int |S(\omega)|^2 d\omega = E \tag{2.41}$$

Then, the normalized functions $|s(t)|^2/E$ and $|S(\omega)|^2/2\pi E$ can be thought of as the signal energy density functions in the time and frequency domains, respectively. In this case, we can use the moment concepts from the probability theory to quantitatively characterize the signal's behaviors. For example, we could use the first moment to compute the mean time and mean frequency, i.e.,

$$\langle t \rangle = \frac{1}{E}\int t|s(t)|^2 dt \tag{2.42}$$

and

$$\langle \omega \rangle = \frac{1}{2\pi E}\int \omega|S(\omega)|^2 d\omega \tag{2.43}$$

Without loss of the generality, let $s(t) = A(t)\exp\{j\varphi(t)\}$, where $A(t)$ and $\varphi(t)$ are magnitude and phase, respectively. Both of them are real.

Eq.(2.43) implies that to calculate the mean frequency, we have to first compute Fourier transform. In what follows, we shall derive a formula to compute the mean frequency without calculating Fourier transform. The result not only simplifies the procedure of computing the mean frequency, but also helps us to better understand the relationship between time representation and frequency behavior.

Let $H(\omega) = \omega S(\omega)$, then by derivative property, we have

$$h(t) = -j\frac{d}{dt}s(t) \tag{2.44}$$

Applying Parseval's formula (2.39) on (2.43) obtains,

$$
\begin{aligned}
<\omega> &= \frac{1}{2\pi E}\int H(\omega)S^*(\omega)d\omega \\
&= \frac{1}{E}\int -j\frac{d}{dt}s(t)s^*(t)dt \\
&= \frac{1}{E}\int -j\big(A'(t)\exp\{j\varphi(t)\} + A(t)j\varphi'(t)\exp\{j\varphi(t)\}\big)s^*(t)dt \\
&= \frac{1}{E}\int -j\big(A'(t) + A(t)j\varphi'(t)\big)A(t)dt \\
&= \frac{1}{E}\int \varphi'(t)[A(t)]^2\,dt - \frac{j}{E}\int A'(t)A(t)dt
\end{aligned}
\tag{2.45}
$$

Because the left side of (2.45), the mean frequency, is real, the second term of the right side in (2.45) must be zero. Consequently, (2.45) reduces to

$$\langle\omega\rangle = \int\varphi'(t)\frac{|s(t)|^2}{E}dt \tag{2.46}$$

which says that the mean frequency $<\omega>$ is the weighted average of the instantaneous quantities $\varphi'(t)$ over entire time domain. We name $\varphi'(t)$ the *mean instantaneous frequency*.

Traditionally, the first derivative of phase $\varphi'(t)$ is defined as the instantaneous frequency rather than the *mean instantaneous frequency* as we do here. There are two reasons to name $\varphi'(t)$ the *mean instantaneous frequency* rather than the instantaneous frequency. First, at any time instant t, the signal often contains more than one frequency tone, such as the speech signal in Fig. 2–9. Second, each frequency tone has a certain bandwidth. Unless for very special cases, such as complex sinusoidal functions and the constant magnitude linear chirp, the signal spreads out in frequency domain. In other words, the "instantaneous frequency bandwidth" of the signal in general is not equal to zero. The instantaneous frequencies are not the single value function of t. Hence, it is not correct to name $\varphi'(t)$ as the instantaneous frequency. As we shall show in Chapter 5, $\varphi'(t)$ in fact represents the mean or average of the signal's frequencies at time t.

$\varphi'(t)$ is a very important quantity. In joint time-frequency analysis, we often use it to evaluate the merit of a proposed time-dependent spectrum. For a desired time-dependent spectrum $P(t,\omega)$, it is natural that the conditional mean frequency $<\omega>_t$ is equal to the mean instantaneous frequency $\varphi'(t)$. That is,

$$\langle \omega \rangle_t = \frac{\int \omega P(t, \omega) d\omega}{\int P(t, \omega) d\omega} = \varphi'(t) \tag{2.47}$$

When $|s(t)|^2/E$ and $|S(\omega)|^2/2\pi E$ are considered as the signal density functions in the time and frequency domains, we can further use the concept of the variance to measure the signal's energy spreading in time and frequency domains. Usually, we define $2\Delta_t$ and $2\Delta_\omega$ for the time duration and frequency bandwidth, where

$$\Delta_t^2 = \frac{1}{E} \int (t - \langle t \rangle)^2 |s(t)|^2 dt = \int t^2 \frac{|s(t)|^2}{E} dt - \langle t \rangle^2 \tag{2.48}$$

and

$$\Delta_\omega^2 = \frac{1}{2\pi} \int (\omega - \langle \omega \rangle)^2 \frac{|S(\omega)|^2}{E} d\omega = \frac{1}{2\pi} \int \omega^2 \frac{|S(\omega)|^2}{E} d\omega - \langle \omega \rangle^2 \tag{2.49}$$

Eqs.(2.48) and (2.49) are the standard definitions of the variance, which depict the signal's spreading with respect to $<t>$ and $<\omega>$, respectively. Like the case of mean frequency discussed early, we can also express the frequency variance (2.49) as the function of time.

Based on Parseval's relationship, we can rewrite (2.49) as

$$\Delta_\omega^2 = \frac{1}{2\pi E} \int (\omega - \langle \omega \rangle)^2 |S(\omega)|^2 d\omega = \frac{1}{2\pi E} \int H(\omega) H^*(\omega) d\omega = \frac{1}{E} \int h(t) h^*(t) dt \tag{2.50}$$

where $H(\omega) = (\omega - <\omega>)S(\omega)$. Because

$$\frac{1}{2\pi} \int S(\omega) \exp\{j(\omega - \langle \omega \rangle) t\} d\omega = s(t) \exp\{-j\langle \omega \rangle t\} \tag{2.51}$$

Taking the derivative in both sides of (2.51) yields

$$\frac{1}{2\pi} \int j(\omega - \langle \omega \rangle) S(\omega) \exp\{j(\omega - \langle \omega \rangle) t\} d\omega = \frac{d}{dt}[s(t) \exp\{-j\langle \omega \rangle t\}] \tag{2.52}$$

Therefore,

$$
\begin{aligned}
h(t) &= \frac{1}{2\pi}\int H(\omega)\exp\{j\omega t\}d\omega = \frac{1}{2\pi}\int (\omega - \langle\omega\rangle)S(\omega)\exp\{j\omega t\}d\omega \\
&= -je^{j\langle\omega\rangle t}\frac{d}{dt}s(t)\exp\{-j\langle\omega\rangle t\} \\
&= -je^{j\langle\omega\rangle t}[j(\varphi'(t) - \langle\omega\rangle)A(t)\exp\{j(\varphi(t) - \langle\omega\rangle)t\} + A'(t)\exp\{j(\varphi(t) - \langle\omega\rangle)t\}] \\
&= \exp\{j\varphi(t)\}[(\varphi'(t) - \langle\omega\rangle)A(t) - jA'(t)]
\end{aligned}
$$

Substituting $h(t)$ into (2.50) obtains

$$
\Delta_\omega^2 = \frac{1}{E}\int (\varphi'(t) - \langle\omega\rangle)^2 A^2(t)dt + \frac{1}{E}\int (A'(t))^2 dt \tag{2.53}
$$

which says that the frequency bandwidth is completely determined by magnitude variation $A'(t)$ as well as phase variation $\varphi'(t)$. To obtain a narrow band signal, we could either smooth the magnitude or phase. If both the magnitude and phase are constant, such as the complex sinusoidal signal $\exp\{j\omega_0 t\}$, then the frequency bandwidth reduces to zero. To better understand this important concept, let's look at some examples.

Example 2-6   Normalized Gaussian function

$$
s(t) = g(t) = \left(\frac{\alpha}{\pi}\right)^{\frac{1}{4}}\exp\left\{-\frac{\alpha}{2}t^2\right\} \tag{2.54}
$$

Obviously,

$$
E = \int |s(t)|^2 dt = 1
$$

and $\varphi'(t) = 0$. Based on formula (2.46), the mean frequency $\langle\omega\rangle = 0$. From formula (2.53), the variance is

$$
\Delta_\omega^2 = \alpha^2\sqrt{\frac{\alpha}{\pi}}\int t^2\exp\{-\alpha t^2\}dt = \frac{\alpha}{2} \tag{2.55}
$$

Example 2-7   Frequency modulated Gaussian function

$$
s(t) = g(t)\exp\{j\omega_0 t\} \tag{2.56}
$$

then,

$$\varphi'(t) = \omega_0$$

$$\langle\omega\rangle = \sqrt{\frac{\alpha}{\pi}}\int\omega_0\exp\{-\alpha t^2\}dt = \omega_0$$

$$\Delta_\omega^2 = \alpha^2\sqrt{\frac{\alpha}{\pi}}\int t^2\exp\{-\alpha t^2\}dt = \frac{\alpha}{2}$$

Compared to Example 2–6, although the phase in this example is not equal to zero, the variation of the phase is constant, that is, $\varphi'(t) = \omega_0$. Consequently, the frequency bandwidth is unchanged.

Example 2-8   Linear chirp signal with the Gaussian envelope

If

$$s(t) = g(t)\exp\{j\beta t^2\} \tag{2.57}$$

then,

$$\varphi'(t) = 2\beta t$$

$$\langle\omega\rangle = \sqrt{\frac{\alpha}{\pi}}\int 2\beta t\exp\{-\alpha t^2\}dt = 0$$

$$\Delta_\omega^2 = \alpha^2\sqrt{\frac{\alpha}{\pi}}\int t^2\exp\{-\alpha t^2\}dt + \sqrt{\frac{\alpha}{\pi}}\int(2\beta t)^2\exp\{-\alpha t^2\}dt = \frac{\alpha}{2} + \frac{4\beta^2}{\alpha}$$

Because the derivation of the phase in (2.57) is not constant, that is, $\varphi'(t) = 2\beta t$, the frequency bandwidth in this example is larger than that in either Example 2–6 or Example 2–7. The extra term $4\beta^2/\alpha$ is proportional to the sweeping rate β.

Example 2-9   Scaling function $s(t/a)$

In this case, the signal energy is

$$\int|s(t/a)|^2 dt = a\int|s(t/a)|^2 d(t/a) = aE \tag{2.58}$$

The first derivatives of the magnitude and phase are $a^{-1}A'(t/a)$ and $a^{-1}\varphi'(t/a)$, respectively. If the mean frequency and frequency bandwidth of $s(t)$ are $\langle\omega\rangle$ and $2\Delta_\omega$, then for scaled function $s(t/a)$,

$$\langle\omega(a)\rangle = \frac{1}{aE}\int a^{-1}\varphi'(t/a)A^2(t/a)dt = \frac{1}{aE}\int\varphi'(t/a)A^2(t/a)d(t/a) = \frac{\langle\omega\rangle}{a}$$

$$\Delta_\omega^2(a) = \frac{1}{aE}\int\{[a^{-1}A(t/a)]^2 + (a^{-1}\varphi'(t/a) - \langle\omega(a)\rangle)^2 A^2(t/a)\}dt$$

$$= \frac{1}{a^2 E}\int\{A^2(t/a) + (\varphi'(t/a) - a\langle\omega\rangle)^2 A^2(t/a)\}d(t/a)$$

$$= \frac{1}{a^2}\Delta_\omega^2$$

which indicates that time scaling leads to the mean frequency shifting and the frequency bandwidth scaling as illustrated in Fig. 2–10. It is interesting to note, however, the ratio between the bandwidth and the mean frequency is independent of the scaling fact a, that is

$$\frac{2\Delta_\omega(a)}{\langle\omega(a)\rangle} = \frac{2\Delta_\omega}{\langle\omega\rangle} = Q \tag{2.59}$$

In other words, the scaling does not change the ratio between the frequency bandwidth and mean frequency. This property is commonly referred to as the *constant Q*.

Applying the moment concepts, we quantitatively characterize the signal in time and frequency domains, respectively, which greatly facilitates the signal analysis. It should be kept in mind, however, that the moment concepts in general do not directly apply to the joint time-frequency domain. This is because the time-dependent spectrum may go to negative. Consequently, the instantaneous frequency bandwidth, the conditional variance, may become less than zero, which is contradictory to our intuitions. We shall discuss this issue in more detail later.

2.5 Uncertainty Principle

Based on the definitions of time duration and frequency bandwidth given in the previous section, we are now ready to give the quantitative definition of the uncertainty principle as follows.

THEOREM

If

$$\sqrt{t}s(t) \to 0 \tag{2.60}$$

for $|t| \to \infty$, then

$$\Delta_t \Delta_\omega \geq \frac{1}{2} \tag{2.61}$$

The equality only holds when $s(t)$ is the Gaussian function, i.e.,

$$s(t) = A \exp\{-\alpha t^2\} \tag{2.62}$$

PROOF

For the sake of simplicity, let's assume that $<t> = 0$ and $<\omega> = 0$. Consequently, (2.48) and (2.49) become

$$\Delta_t^2 = \int t^2 |s(t)|^2 dt \tag{2.63}$$

and

$$\Delta_\omega^2 = \frac{1}{2\pi} \int \omega^2 |S(\omega)|^2 d\omega \tag{2.64}$$

Then,

$$\Delta_t^2 \Delta_\omega^2 = \int t^2 |s(t)|^2 dt \frac{1}{2\pi} \int \omega^2 |S(\omega)|^2 d\omega \tag{2.65}$$

Replacing $\omega S(\omega) = H(\omega)$ in (2.65) yields

$$\Delta_t^2 \Delta_\omega^2 = \int t^2 |s(t)|^2 dt \int h(t) h^*(t) dt \tag{2.66}$$

where we applied Parseval's relation. Because of the derivative property in (2.34),

$$\omega S(\omega) \leftrightarrow -j\frac{d}{dt} s(t) \tag{2.67}$$

Applying (2.67) to (2.66) yields

$$\Delta_t^2 \Delta_\omega^2 = \int t^2 |s(t)|^2 dt \int \left|\frac{d}{dt} s(t)\right|^2 dt \tag{2.68}$$

From Schwarz' inequality, it follows that

$$\int t^2 |s(t)|^2 dt \int \left|\frac{d}{dt} s(t)\right|^2 dt \geq \left|\int ts(t)\frac{d}{dt} s(t) dt\right|^2 \tag{2.69}$$

Because

$$\int ts(t)\frac{d}{dt} s(t) dt = \frac{1}{2}\int t\frac{d}{dt} s^2(t) dt = \left.\frac{ts^2(t)}{2}\right|_{-\infty}^{\infty} - \frac{1}{2}\int_{-\infty}^{\infty} s^2(t) dt = -\frac{1}{2} \tag{2.70}$$

Inserting (2.70) into (2.69), we obtain the uncertainty inequality relation (2.61). If (2.61) is an equality, then (2.69) must also be an equality. This is possible only if $s'(t) = kts(t)$, that is, $s(t)$ is given by (2.62).

2.6   Discrete Poisson-Sum Formula

The Poisson-sum formula is very useful in signal analysis [141]. In this section, we shall investigate the discrete version of the Poisson-sum formula, which plays an important role in deriving the discrete Gabor expansion.

Let $\{\tilde{a}[n]\}$ be a periodic sequence of period $L = \Delta MM$. Then, we define $\{\tilde{b}[n]\}$ as the periodic extension of $\{\tilde{a}[n]\}$, that is

$$\tilde{b}[n] = \sum_{m=0}^{M-1} \tilde{a}[n - m\Delta M] \tag{2.71}$$

Obviously,

$$\tilde{b}[n] = \tilde{b}[n + \Delta M] \tag{2.72}$$

Let's denote $B[k]$ the DFT of $\{\tilde{b}[n]\}$, that is,

$$B[k] = \sum_{n=0}^{\Delta M-1} \tilde{b}[n]W_M^{-nk} = \sum_{n=0}^{\Delta M-1}\left(\sum_{m=0}^{M-1}\tilde{a}[n - m\Delta M]\right)W_M^{-nk} \tag{2.73}$$

Substituting $i = n - m\Delta M$, (2.73) reduces to a single summation, such as

$$B[k] = \sum_{i=0}^{L-1} \tilde{a}[i]W_{\Delta M}^{-ik} \tag{2.74}$$

Based on (2.73), we have

$$\tilde{b}[n] = \frac{1}{\Delta M}\sum_{k=0}^{\Delta M-1} B[k]W_{\Delta M}^{nk} = \frac{1}{\Delta M}\sum_{k=0}^{\Delta M-1}\left(\sum_{i=0}^{L-1}\tilde{a}[i]W_{\Delta M}^{-ik}\right)W_{\Delta M}^{nk} \tag{2.75}$$

By (2.71), we have

$$\sum_{m=0}^{M-1}\tilde{a}[n - m\Delta M] = \frac{1}{\Delta M}\sum_{k=0}^{\Delta M-1}\left(\sum_{i=0}^{L-1}\tilde{a}[i]W_{\Delta M}^{-ik}\right)W_{\Delta M}^{nk} \tag{2.76}$$

which is known as the *Poisson-sum formula*.

Summary

From the mathematics point of view, a given signal can be represented in an infinite number of ways via different *expansions*. A typical goal in signal processing is to find a representation in which certain attributes of the signal are made explicit. Therefore, the central issues of signal processing are how to construct a set of the elementary functions $\{\psi_n\}_{n \in Z}$ and how to compute the corresponding dual functions. It is worthwhile to note that $\{\psi_n\}_{n \in Z}$ and $\{\hat{\psi}_n\}_{n \in Z}$ are exchangeable. Either of them can be used for analysis functions to compute the expansion coefficients or the transform. The process of computing the coefficients is similar to using a "ruler," constituted by a set of analysis functions, to measure the signal under consideration. For good measuring, the set of the analysis functions should be selected such that the resulting "ruler" is easier to be built and its tick marks are finest.

Because *time* and *frequency* are the two most fundamental quantities, the time and frequency representations are the most important signal representations. While a signal's time representations are natural, the frequency representations were not popular until the early 19th century, when Fourier first proposed harmonic trigonometric series. The basic method used for exploring a signal's periodicity property is to compare the analyzed signals with the harmonically related sinusoidal functions, which is named the inner product in mathematical literature. The two most common approaches to describe a signal's behavior in terms of frequency are the Fourier transform (linear) and power spectrum (quadratic).

In this section, we reviewed the relationships between time and frequency representations. In particular, we discussed the frequency bandwidth. It is shown that a signal's frequency bandwidth (or frequency resolution) is completely determined by magnitude and phase variations. Because the time and frequency representations are related by the Fourier transformation, a signal's behaviors in the time domain and frequency domain are not independent. For example, we cannot make the signal's time-duration and frequency bandwidth arbitrarily small simultaneously. The product of the time-duration and frequency bandwidth has to satisfy the uncertainty principle.

Joint Time-Frequency Analysis

*T**he Fourier transform has been the most common tool to study a signal's frequency properties. However, based on the Fourier transform and power spectrum alone, it is hard to tell whether or not a signal's frequency contents evolve in time, even though the phase of the Fourier transform relates to time shifting. On the other hand, except for a few special cases, the frequency contents of the majority of signals encountered in the real world change with time. In those applications, the classical Fourier analysis is no longer adequate. As shown in Chapter 2, a good example is the speech signal. Intuitively, the formant of the speech must be time-varying. Otherwise, the speech will be indistinguishable and thereby cannot be used for our daily communications. It has been recognized for a long time that the conventional power spectrum is not suitable for the study of speech, Doppler frequency, as well as many other signals, in which the spectra evolve with time.*

It is the primary goal of this book to introduce the signal's joint time-frequency representations and their applications. Analogous to the classical Fourier analysis, in this part we present in parallel the methods of linear and quadratic (or bilinear) joint time-frequency representations. Chapters 3 and 4 are devoted to the short-time Fourier transform (STFT) and the wavelets transform, respectively. Although results of the STFT and the wavelet transform look quite different, the techniques used to measure the signal's local behavior are identical. All linear transformations are achieved by comparing the analyzed signal with a set of prudently selected elementary functions that can be thought of as the tick marks of the ruler used in our everyday life. The main difference lies in the way in which the tick marks are built. While the frequency tick marks are obtained by frequency modulation in the STFT, in the wavelet transform the frequency tick marks are obtained by scaling the center frequency of the mother wavelet (which is equivalent to reciprocal scaling of the mother wavelet in the time domain).

Unlike the linear representations, which are clearly predominated by the STFT and wavelet transform, there are more than a dozen candidates for the bilinear time-frequency representations. Our discussions, however, start with the Wigner-Ville distribution in Chapter 5, because it is simple and it better characterizes the signal's time-dependent spectra than the STFT spectrogram (square of the short-time Fourier transform), scalogram (square of the wavelets), as well as many other methods known so far. The main deficiency of the Wigner-Ville distribution is the so-called cross-term interference that significantly obscures the applications of the Wigner-Ville distribution. It has long been recognized, however, that the cross-terms always occur in the midway of two auto-terms. They are localized and highly oscillated. On the other hand, the useful properties possessed by the Wigner-Ville distribution are obtained by averaging the Wigner-Ville distribution. Those observations suggest that we may apply a lowpass filter for the Wigner-Ville distribution, to retain the low frequency components and remove the high frequency parts. Because the discarded high oscillated parts have small averages, the lowpass filtered Wigner-Ville distribution presumably preserves the useful properties with reduced cross-term interference. Basically, there are two types of lowpass filters: linear, characterized by Cohen's class in Chapter 6, and non-linear, described by the time-frequency distribution series (also known as the Gabor spectrogram) in Chapter 7. In Chapter 8, we further introduce the signal's adaptive representation and adaptive spectrogram. While the two-dimensional elementary functions have a fixed envelope in the time-frequency distribution series, for the adaptive spectrogram, the elementary functions are adapted to best match the signal under consideration.

In addition to studying the signal's frequency content changes, another major advantage to displaying time function in joint time-frequency domain is for noise reduction. Generally speaking, the noise tends to evenly spread into the entire joint time-frequency domain. In contrast, the signal is concentrated in the relatively small area. As a result, the regional signal-to-noise ratio of noise-corrupted signals could be significantly improved. In general, it is much easier to recognize the noisy signal in the joint time-frequency domain than from either time or frequency domain alone. Once the signal is identified, we then can remove all noise, simply by forcing the noise terms to zero, to obtain a modified noiseless joint time-frequency representation. By the inverse transform, finally the original time signal is recovered. Such an operation is traditionally considered as time-variant filtering and has been found very powerful for the wideband and non-stationary signal estimation.

In principle, all joint time-frequency representations, both linear as well as bilinear, could be used for the time-variant filters. Our discussions in Chapter 9, however, are focused on the Gabor expansion-based time-variant filters, because they are simple and powerful.

The literature on joint time-frequency analysis is enormous, but most of it requires a level of mathematical preparation which is perhaps not suitable for engineering students as well as practicing engineers/scientists. We have tried to present the fundamental ideas and important algorithms only with the help of elementary calculus and linear algebra. All the algorithms introduced in Part 2 have been extensively tested by the authors and National Instruments customers over the last three years. We sincerely hope that the reader will find that the materials in this part are interesting and enlightening.

CHAPTER 3

Short-Time Fourier Transform and Gabor Expansion

*I*n conventional Fourier transform, the signal is compared to complex sinusoidal functions. Because sinusoidal basis functions spread into the entire time domain and are not concentrated in time, the Fourier transform does not explicitly indicate how a signal's frequency contents evolve in time.

Based on the expansion and inner product concepts, a natural way of characterizing a signal in time and frequency simultaneously is to compare the signal with elementary functions that are concentrated in both time and frequency domains, such as the frequency modulated Gaussian function. Because Gaussian-type functions are optimally concentrated in the joint time and frequency domains, the resulting comparisons reflect a signal's behavior in local time and frequency.

In Section 3.1, we briefly introduce the methodology of the short-time Fourier transform[*] (STFT). For the continuous-time STFT, the analysis function and synthesis function have the same form. We can easily recover the original time functions based on the STFT. However, the representation based on the continuous-time STFT is highly redundant (or oversampled). For a compact presentation, we often prefer to use the sampled STFT. In this case, the inverse problem is no longer as straightforward as the case of the continuous-time STFT. The inverse of sampled STFT can be accomplished by the Gabor expansion. Although

---

[*] It is also known as *windowed Fourier transform*.

Gabor was apparently not motivated to investigate the inverse problem of sampled STFT, the Gabor expansion turns out to be the most elegant algorithm of computing the inverse of the sampled STFT.

Section 3.2 is devoted to the general introduction of the Gabor expansion. Although the idea of the Gabor expansion was rather straightforward, its implementation has been a hot research topic. The continuous-time Gabor expansion in fact has a wider scope for deeper mathematical issues, which has been thoroughly studied by Janssen (see [86], [91], [94], and [95]) as well as many other researches. The discrete-time Gabor expansion, on the other hand, is relatively simple and can be realized with the help of elementary linear algebra. In Section 3.3, we introduce the discrete Gabor expansion for periodic sequences. Then, we develop the discrete Gabor expansion for infinite samples in Section 3.4. In general, given the synthesis function, the dual function is not unique. The natural question is how to choose the dual functions. In Section 3.5, we introduce the orthogonal-like Gabor expansion. The concept of the orthogonal-like has been found very important from both a theoretical and application point of view. In Section 3.6, we present a fast algorithm of computing dual functions. In appendix A, we discuss the existence of the biorthogonal function at critical sampling. In appendix B, we investigate the general optimal algorithm that allows the dual function $\gamma[k]$ to be optimally close to an arbitrary desired function $d[k]$.

Over the years, many techniques have been successfully developed to implement the Gabor expansion, such as Zak transform-based algorithms (see [198] and [199]), filter bank methods [179], as well as the pseudo-frame approach [115]. In this book, we have limited our discussions to the method that was first introduced by Bastiaans (see [8], [9], and [10]) and recently extended by Wexler and Raz [185]. The reader who is interested in methods other than those presented in this chapter may consult the related literature.

3.1 Short-Time Fourier Transform

The frequency contents of the majority of signals encountered in our everyday life change over time, such as biomedical signals, speech signals, stock indexes, and vibrations. Because the basis functions used in the classical Fourier analysis do not associate with any particular time instant, the resulting measurements, Fourier transforms, do not explicitly reflect a signal's time-varying nature[*].

A simple way to overcome the deficiency possessed by the regular Fourier transform is to compare the signal with elementary functions that are localized in time and frequency domains simultaneously, i.e.,

$$\mathrm{STFT}(t, \omega) = \int s(\tau)\gamma^*_{t,\omega}(\tau)d\tau = \int s(\tau)\gamma^*(\tau - t)e^{-j\omega\tau}d\tau \qquad (3.1)$$

---

[*] Although the phase characteristic of $S(\omega)$ contains the time information, it is difficult to establish the point-to-point relationship between $s(t)$ and $S(\omega)$ based upon the conventional Fourier analysis.

which is a regular inner product and reflects the similarity between signal $s(t)$ and the elementary function $\gamma(\tau{-}t)\exp\{j\omega\tau\}$. The function $\gamma(t)$ usually has a short time duration and thereby it is named the *window function*. Eq. (3.1) is called *short-time Fourier transform* (STFT) or windowed Fourier transform.

The formula (3.1) can be understood in several ways. Fig. 3–1 depicts the procedure of computing the STFT; first multiply the function $\gamma(t)$ with signal $s(t)$ and compute the Fourier transform of the product $s(\tau)\gamma*(\tau{-}t)$. Because the window function $\gamma(t)$ has a short time duration, the Fourier transform of $s(\tau)\gamma*(\tau{-}t)$ reflects the signal's local frequency properties. By moving $\gamma(t)$ and repeating the same process, we could obtain a rough idea how the signal's frequency contents evolve over time.

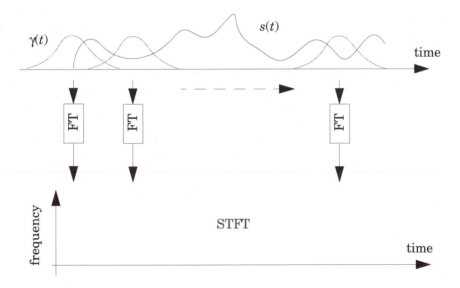

Fig. 3–1 Short-Time Fourier Transform

Alternatively, we could also understand STFT from the concept of expansion introduced in Chapter 2. In STFT, we compare the signal $s(t)$ with a set of elementary functions $\gamma(\tau{-}t)\exp\{j\omega\tau\}$ that are concentrated in both time and frequency domains. Suppose that the function $\gamma(t)$ is centered at $t = 0$ and its Fourier transform is centered at $\omega = 0$. If the time duration and frequency bandwidth of $\gamma(t)$ are Δ_t and Δ_ω, then STFT(t,ω) in (3.1) indicates a signal's behavior in the vicinity of $[t{-}\Delta_t,\ t{+}\Delta_t] \times [\omega{-}\Delta_\omega,\ \omega{+}\Delta_\omega]$.

In order to better measure a signal at a particular time and frequency (t,ω), it is natural to desire that Δ_t and Δ_ω be as narrow as possible. Unfortunately, the selections of Δ_t and Δ_ω are not independent, which are related via the Fourier transform. If we let Δ_t and Δ_ω be a signal's standard deviations as introduced in

Chapter 2, then the product $\Delta_t \Delta_\omega$ has to satisfy the uncertainty inequality, that is

$$\Delta_t \Delta_\omega \geq \frac{1}{2}$$

There is a trade-off of the selection of the time and frequency resolution. If $\gamma(t)$ is chosen to have good time resolution (smaller Δ_t), then its frequency resolution must be deteriorated (larger Δ_ω), or vice versa. The equality only holds when $\gamma(t)$ is a Gaussian function.

The reader should bear in mind that if $\{\gamma(\tau-t)\exp\{j\omega\tau\}\}$ is considered as a ruler, then the different time and frequency tick marks for STFT are obtained by time-shifting and frequency-modulating a single prototype window function $\gamma(t)$. In the next chapter, we shall introduce another way to constitute time and frequency tick marks, which is commonly known as *wavelets*. Although the concepts of exploring a signal's time-varying nature are quite similar (performing inner product operations), the different way of building tick marks leads to very different outcomes.

The square of STFT is named STFT spectrogram to distinguish it from the time-dependent spectrum based upon other linear techniques, such as the Gabor expansion and the adaptive representations. STFT spectrogram is the most simple and used time-dependent spectrum, which roughly depicts a signal's energy distribution in the joint time-frequency domain. While the STFT in general is complex, the STFT spectrogram is always real-valued.

Example 3–1   STFT with Gaussian-type analysis
             function

If

$$\gamma(t) = \left(\frac{\alpha}{\pi}\right)^{\frac{1}{4}} \exp\left\{-\frac{\alpha}{2}t^2\right\} \qquad (3.2)$$

and

$$s(t) = \left(\frac{\beta}{\pi}\right)^{\frac{1}{4}} \exp\left\{-\frac{\beta}{2}t^2\right\} \qquad (3.3)$$

Intuitively, in the joint time-frequency domain, $s(t)$ is centered in (0,0). Its time duration is determined by β. Substituting $s(t)$ and $\gamma(t)$ into (3.1) yields

$$\text{STFT}(t, \omega) = \left(\frac{\alpha\beta}{\pi^2}\right)^{\frac{1}{4}} \int \exp\left\{-\frac{\beta}{2}\tau^2\right\} \exp\left\{-\frac{\alpha}{2}(\tau - t)^2\right\} \exp\{-j\omega\tau\} d\tau$$

$$= \left(\frac{\alpha\beta}{\pi^2}\right)^{\frac{1}{4}} \int \exp\left\{-\left(\frac{\alpha+\beta}{2}\right)\tau^2 + \alpha t\tau - \frac{\alpha}{2}t^2\right\} \exp\{-j\omega\tau\} d\tau$$

$$= \exp\left\{-\frac{\alpha\beta}{2(\alpha+\beta)}t^2\right\} \left(\frac{\alpha\beta}{\pi^2}\right)^{\frac{1}{4}} \int \exp\left\{-\left(\frac{\alpha+\beta}{2}\right)\left[\tau^2 - \frac{2\alpha t}{\alpha+\beta}\tau + \left(\frac{\alpha t}{\alpha+\beta}\right)^2\right]\right\} \exp\{-j\omega\tau\} d\tau$$

$$= \exp\left\{-\frac{\alpha\beta}{2(\alpha+\beta)}t^2\right\} \left(\frac{\alpha\beta}{\pi^2}\right)^{\frac{1}{4}} \int \exp\left\{-\left(\frac{\alpha+\beta}{2}\right)\left(\tau - \frac{\alpha}{\alpha+\beta}t\right)^2\right\} \exp\{-j\omega\tau\} d\tau$$

Applying the *Gaussian characteristic function* introduced in Example 2–4, we have

$$\text{STFT}(t, \omega) = \left(\frac{2\sqrt{\alpha\beta}}{\alpha+\beta}\right)^{\frac{1}{2}} \exp\left\{-\frac{\alpha\beta}{2(\alpha+\beta)}t^2 - \frac{1}{2(\alpha+\beta)}\omega^2 + j\frac{\alpha}{\alpha+\beta}\omega t\right\} \qquad (3.4)$$

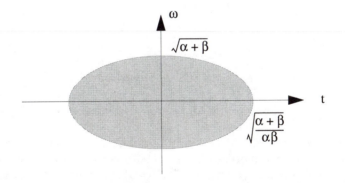

Fig. 3–2 The ellipse area reaches its minimum when the variance of analysis function perfectly matches the signal time duration, that is, $\alpha = \beta$. The minimum area is 2π, which is twice as large as that of the Wigner-Ville distribution.

The corresponding STFT spectrogram is

$$\text{SP}(t, \omega) = |\text{STFT}(t, \omega)|^2 = \frac{2\sqrt{\alpha\beta}}{\alpha+b} \exp\left\{-\frac{\alpha\beta}{\alpha+\beta}t^2 - \frac{1}{\alpha+\beta}\omega^2\right\} \qquad (3.5)$$

which shows that STFT spectrogram is concentrated in (0,0), the center of signal $s(t)$. The contours of equal height of SP in (3.5) are ellipses. The contour for the case where the levels are down to e^{-1} of their peak value is the ellipse indicated in Fig. 3–2. The area of this particular level ellipse is

$$A = \frac{\alpha + \beta}{\sqrt{\alpha\beta}}\pi = \frac{1 + r}{\sqrt{r}}\pi \tag{3.6}$$

where $r = \beta/\alpha$ is a matching indicator. The area A reflects the concentration of STFT. Naturally, the smaller the A is, the better the resolution. The resolution of STFT is subject to the selection of analysis function. The minimum of A in (3.6) occurs when $r = 1$. In other words, when the variance of analysis function α perfectly matches the time duration of the analyzed signal β, $SP(t,\omega)$ in (3.5) will have the best resolution. However, because the signal duration β will likely be unknown, it would be difficult to achieve the optimal resolution in general. Moreover, it should bear in mind that even the optimal resolution, $A = 2\pi$, is twice as large as that of the Wigner-Ville distribution.

Taking the inverse Fourier transform with respect to STFT(t,ω) in (3.1) yields

$$\frac{1}{2\pi}\int \text{STFT}(t, \omega)\exp\{j\mu\omega\}d\omega = \frac{1}{2\pi}\iint s(\tau)\gamma(\tau - t)\exp\{j(\mu - \tau)\omega\}d\tau d\omega$$

$$= \int s(\tau)\gamma(\tau - t)\delta(\mu - \tau)d\tau$$

$$= s(\mu)\gamma(\mu - t)$$

Let $\mu = t$, we have

$$s(t) = \frac{1}{2\pi\gamma(0)}\int \text{STFT}(t, \omega)\exp\{jt\omega\}d\omega \tag{3.7}$$

which implies given STFT(t,ω) for all t and ω, we can completely recover the signal $s(t)$.

It is worth noting that (3.7) is a highly redundant representation. In fact, the signal $s(t)$ can be completely reconstructed merely from the sampled version of the short-time Fourier transform, STFT$(mT,n\Omega)$, where T and Ω denote the time and frequency sampling steps, respectively. In other words, we can use the sampled STFT to completely characterize signal $s(t)$ and thereby save considerable computation as well as memory. Unfortunately, in this case, the reconstruction is no longer as simple as (3.7). We shall discuss this subject in more detail in the subsequent sections.

The STFT can also be viewed as a mapping from the time domain to the time-frequency domain, as illustrated in Fig. 3–3. For any time domain function $s(t)$ and window function $\gamma(t)$, such mapping always exists. But the inverse may not be true. In other words, given a function $\gamma(t)$ and an arbitrary two-dimensional function $B(t,\omega)$, there may be no physically existing signals $s(t)$ whose

STFT is equal to $B(t, \omega)$. In this case, we say that $B(t, \omega)$ is not a valid short-time Fourier transform. The simplest example is

$$B(t, \omega) = \begin{cases} 1 & \text{for } |t| < t_0 |\omega| < \omega_0 \\ 0 & \text{otherwise} \end{cases} \tag{3.8}$$

Because no signal can be finite supported in both time and frequency domains, $B(t, \omega)$ cannot be a valid joint time-frequency representation. As shown in Fig. 3–3, STFT(t, ω) in fact is only a subset of the entire two-dimensional function $B(t, \omega)$.

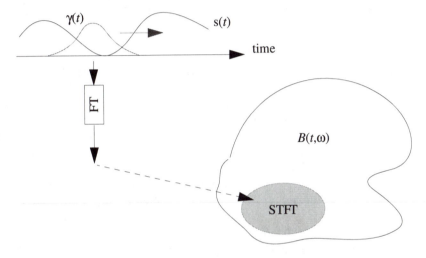

Fig. 3–3 STFT(t, ω) is a subset of the entire two-dimensional function. An arbitrary two-dimensional time-frequency function may not be a valid STFT(t, ω).

For a given function $\gamma(t)$, a valid short-time Fourier transform has to be such that its inverse Fourier transform is separable, that is

$$\frac{1}{2\pi} \int B(t, \omega) \exp\{j\mu\omega\} d\omega = s(\mu)\gamma(\mu - t) \qquad \forall\, t, \mu \tag{3.9}$$

For the digital signal processing application, it is necessary to extend the STFT framework to discrete-time signal. For the practical implementation, each Fourier transform in the STFT has to be replaced by the discrete Fourier transform, the resulting STFT is discrete in both time and frequency and thus is suitable for digital implementation, i.e.,

$$\text{STFT}[k, n] = \sum_{i=0}^{L-1} s[i]\gamma[i-k]W_L^{-ni} \tag{3.10}$$

where

$$\mathrm{STFT}[k, n] \equiv \mathrm{STFT}(t, \omega)\big|_{t = k\Delta t, \, \omega = 2\pi n/(L\Delta t)}$$

where Δt denotes the time sampling interval. $\gamma[k] \equiv \gamma(k\Delta t)$ is the L-point window function. We call (3.10) the discrete STFT to distinguish it from the discrete-time STFT, which is continuous in frequency. It is rather easy to verify that the discrete STFT is periodic in frequency, that is,

$$\mathrm{STFT}[k, n] = \mathrm{STFT}[k, n + lL]$$

for $l = 0, \pm 1, \pm 2, \pm 3 \dots$. Like the continuous-time STFT, arbitrary two-dimensional discrete function in general is not a valid discrete short-time Fourier transform.

3.2   Gabor Expansion - Inverse Sampled STFT

Instead of representing a signal either as the function of time or the function of frequency separately, in 1946, Gabor[*] suggested representing a signal in two dimensions, with time and frequency as coordinates [53]. Gabor named such two-dimensional representations the "information diagrams as areas in them are proportional to the number of independent data which they can convey. Gabor pointed out that there are certain "elementary signals" which occupy the smallest possible area in the information diagram. Each elementary signal can be considered as conveying exactly one datum, or one "quantum of information." Any signal can be expanded in terms of these by a process which includes time analysis and frequency analysis as extreme cases. For signal $s(t)$, the Gabor expansion is defined as

$$s(t) = \sum_{m=-\infty}^{\infty} \sum_{n=-\infty}^{\infty} C_{m,n} h_{m,n}(t) = \sum_{m=-\infty}^{\infty} \sum_{n=-\infty}^{\infty} C_{m,n} h(t - mT) \exp\{jn\Omega t\} \qquad (3.11)$$

where T and Ω denote the time and frequency sampling steps. Fig. 3–4 illustrates the Gabor sampling grid.

---

[*] Dennis Gabor was born on June 5, 1900, in Budapest, Hungary. His talent for memorization — an asset in any academic field — appeared at the age of twelve, when he earned a prize from his father for learning by heart, in German, a 430-line poem. Gabor finished his doctorate in electrical engineering in 1927. His work in communication theory and holography started at the end of World War II. It was during that time that he wrote the famous Gabor expansion paper. In 1949, he joined the Imperial College of Science and Technology at London University and in the late sixties became a staff scientist at CBS Laboratories in the United States. While his formal education had been largely in the applied engineering fields, he had not neglected to study the basic physical and mathematical tools that would facilitate his life's work, which was mostly motivated by a desire to create or perfect a particular device invariably secured on a sound mathematical footing. His genius as an inventor lay in an innate ability to focus on a final goal, regardless of the difficulties. His endeavor paid off. In 1971, the Royal Swedish Academy of Sciences presented Dennis Gabor with the Nobel Prize for his discovery of the principles underlying the science of holography.

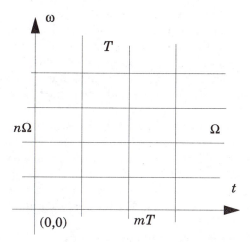

Fig. 3–4 Gabor sampling lattice

In Gabor's original paper, he selected the Gaussian function as the elementary function, i.e.,

$$h(t) = \left(\frac{\alpha}{\pi}\right)^{\frac{1}{4}} \exp\left\{-\frac{\alpha}{2}t^2\right\} \tag{3.12}$$

because the Gaussian function is optimally concentrated in the joint time-frequency domain according to the *uncertainty principle*, that is,

$$\Delta_t \Delta_\omega = \frac{1}{\sqrt{\alpha}}\frac{\sqrt{\alpha}}{2} = \frac{1}{2}$$

which is the lower bound of the uncertainty inequality.

Although Gabor restricted himself to an elementary signal that has a Gaussian shape, his signal expansion in fact holds for rather arbitrarily shaped signals. For almost any signal $h(t)$, its time-shifted and harmonically modulated version can be used as the *Gabor elementary functions*. The necessary condition of the existence of the Gabor expansion is that the sampling cell $T\Omega$ must be small enough to satisfy

$$T\Omega \le 2\pi \tag{3.13}$$

Intuitively, if the sampling cell $T\Omega$ is too large, we may not have enough information to completely recover the original signal. On the other hand, if the sampling cell $T\Omega$ is too small, the representation will be redundant. Traditionally, it is called *critical sampling* when $T\Omega = 2\pi$ and *oversampling* when $T\Omega < 2\pi$. It is interesting to note that although Gabor was not known to have investigated the existence of the formula (3.11), the sampling cell that he selected, $T\Omega = 2\pi$, happened to be the most compact representation!

Gabor was also not known to have published any practical algorithms of computing the Gabor coefficients[*]. Despite the earlier treatment by Auslander et al [3], Gabor's work was not very popular until 1980, in particular, after Bastiaans related the Gabor expansion and short-time Fourier transform (see [8], [9], and [10]). Bastiaans introduced the sampled short-time Fourier transform to compute the Gabor coefficients.

As mentioned in the preceding section, the continuous-time inverse STFT is a highly redundant expansion. In applications, for a compact presentation we usually use sampled STFT. However, the imprudent choice of analysis function $\gamma(t)$ and sampling steps, T and Ω, may lead to the sampled STFT being non-invertible. With the help of the Gabor expansion, we now can easily solve the problem of the inverse of sampled STFT, even though it was apparently not Gabor's original motivation.

Based upon the expansion theorem introduced in Chapter 2, if the set of the Gabor elementary functions $\{h_{m,n}(t)\}$ is complete, then there will be a dual function (or auxiliary function) $\gamma(t)$ such that the Gabor coefficients can be computed by the regular inner product operation, i.e.,

$$C_{m,n} = \int s(t)\gamma^*_{m,n}(t)dt = \int s(t)\gamma^*(t-mT)\exp\{-jn\Omega t\}dt$$
$$= \text{STFT}(mT, n\Omega) \tag{3.14}$$

which is the sampled STFT and also known as the *Gabor transform*. It can be shown that for critical sampling, the Gabor elementary functions $\{h_{m,n}(t)\}$ are linearly independent. In this case, the dual function is unique and biorthogonal to $h(t)$. At oversampling, the selection of the auxiliary function is not unique. There are two fundamental problems regarding the implementation of the Gabor expansion:

- how to compute the dual functions $\gamma(t)$?
- how to select the dual function $\gamma(t)$ if they are not unique?

Substituting (3.14) into the right side of (3.11) yields

$$s(t) = \int s(t') \sum_{m=-\infty}^{\infty} \sum_{n=-\infty}^{\infty} \gamma^*_{m,n}(t')h_{m,n}(t)dt' \tag{3.15}$$

Obviously, the Gabor expansion exists if and only if the double summation is a

---

[*] In his notable paper, Gabor proposed an iteration approach to compute the coefficients $C_{m,n}$, which, however, has been found not to converge in general [54].

delta function, that is,

$$\sum_{m=-\infty}^{\infty} \sum_{n=-\infty}^{\infty} \gamma^*_{m,n}(t') h_{m,n}(t) = \delta(t-t') \tag{3.16}$$

By the Poisson-sum formula, (3.16) can be reduced to a single integration [185], i.e.,

$$\frac{T_0 \Omega_0}{2\pi} \int h(t) \gamma^{0\,*}_{m,n}(t)\, dt = \delta(m)\delta(n) \tag{3.17}$$

where

$$\gamma^0_{m,n} = \gamma(t - mT_0) \exp\{jn\Omega_0\} \tag{3.18}$$

where $T_0 = 2\pi/\Omega$ and $\Omega_0 = 2\pi/T$. In some literature, (3.17) is named the *Wexler-Raz identity,* which plays an important role in computing the dual functions. Note that except for the critical sampling, $T\Omega = 2\pi$, $\gamma_{m,n}(t) \neq \gamma^0_{m,n}(t)$.

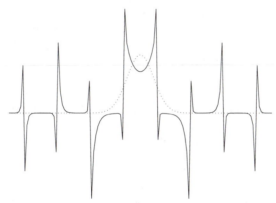

Fig. 3–5 Although the Gabor elementary function (dotted line) is optimally concentrated in the joint time-frequency domain, the corresponding biorthogonal function (solid line) is neither localized in time nor in frequency.

At critical sampling, the set of $\{h_{m,n}(t)\}$ is linearly independent. In this case, we say $\gamma(t)$ and $h(t)$ are biorthogonal to each other. For oversampling, the set of $\{h_{m,n}(t)\}$ is linearly dependent. The resulting presentation is redundant. Bastiaans gave the closed form of the solution $\gamma(t)$ for the Gaussian function $h(t)$ at critical sampling. The results are plotted in Fig. 3–5. Note that, although $h(t)$ (dotted line) is optimally concentrated in the joint time-frequency domain, the corresponding biorthogonal function $\gamma(t)$ (solid line) is neither concentrated in time nor in frequency.

It is worth noting that, unlike harmonically related complex sinusoidal functions used in the Fourier series, which are orthonormal, the set of the Gabor elementary functions in general does not constitute an orthogonal basis. In this case, the dual function $\gamma(t)$ is not equal to the Gabor elementary function $h(t)$. The direct consequence is that although we could easily have the Gabor elementary functions optimally concentrated in the joint time-frequencies domain, the dual function $\gamma(t)$ may not be localized. Consequently, the Gabor coefficients $C_{m,n}$, the inner product of signal and dual functions, do not necessarily reflect the signal's behavior in the vicinity of $[mT-\Delta_t,\ mT+\Delta_t]\times[n\Omega-\Delta_\omega,\ n\Omega+\Delta_\omega]$. Whether or not the Gabor coefficients $C_{m,n}$ describe the signal's local behavior depends on the property of the dual function. If $\gamma(t)$ is badly concentrated in the joint time-frequency domain, the Gabor coefficients $C_{m,n}$ will fail to describe the signal's local behavior. We shall discuss a great deal of the selection of dual functions in Section 3.5.

Finally, we should emphasize that the dual functions $\gamma(t)$ and $h(t)$ are exchangeable. The Gabor expansion can be written in either way as

$$s(t) = \sum_{m,n} \langle s, \gamma_{m,n}\rangle h_{m,n}(t) = \sum_{m,n} \langle s, h_{m,n}\rangle \gamma_{m,n}(t) \qquad (3.19)$$

Which one, $\gamma(t)$ or $h(t)$, is used for the analysis function to compute the Gabor coefficients depends on the applications at hand. If we are mainly interested in Gabor coefficients, then we may use $h_{m,n}(t)$ to calculate $C_{m,n}$, because it is selected first and thereby easier to make it meet our requirements. In this case, once $h(t)$ is properly selected, the Gabor coefficients $C_{m,n}$ will well describe the signal's local time and frequency behaviors.

3.3 Gabor Expansion for Discrete Periodic Samples[*]

The utilization of the Gabor expansion (or the inverse sampled STFT) hinges on the availability of the dual function. Except for a few specific functions, such as Gaussian, two- and one-sided exponential at critical sampling, where the dual functions (in this case, the dual functions must be biorthogonal to each other) can be explicitly computed (see [8], [41], and [52]), analytical solutions of $\gamma(t)$ are not generally available. Moreover, the signals encountered in most applications today are of discrete-time. Hence, it is necessary and beneficial to extend Gabor's framework into the case of discrete-time and discrete-frequency.

The procedure of digitizing the continuous-time Gabor expansion (3.11) essentially is a standard sampling process. Thereby, we leave it for the reader to exercise. Note that sampling the time variables leads to periodicity in the frequency domain. Similarly, digitizing the frequency variable results in periodicity in the time domain. Because it digitizes both time and frequency indices, the discrete version of the Gabor expansion in general is only applied for the periodic discrete-time signals.

---

[*] Because we can easily extend a finite sequence to a periodic function, all results developed from periodic functions are automatically applicable for finite samples.

For discrete-time signal $s[k]$ with period L, the discrete Gabor expansion is defined by

$$\tilde{s}[k] = \sum_{m=0}^{M-1} \sum_{n=0}^{N-1} \tilde{C}_{m,n} \tilde{h}[k - m\Delta M] W_L^{n\Delta Nk} \tag{3.20}$$

where the Gabor coefficients are computed by

$$\tilde{C}_{m,n} = \sum_{k=0}^{L-1} \tilde{s}[k] \tilde{\gamma}^*[k - m\Delta M] W_L^{-n\Delta Nk} \tag{3.21}$$

Note that the signal $s[k]$, synthesis function $h[k]$, and analysis function $\gamma[k]$ are all periodic and have the same period L. We name (3.20) the *periodic discrete Gabor expansion* to distinguish it from the *discrete Gabor expansion* in which neither the analyzed signal nor the window functions need to be periodic. Fig. 3–6 depicts the procedure computing the periodic discrete Gabor coefficients.

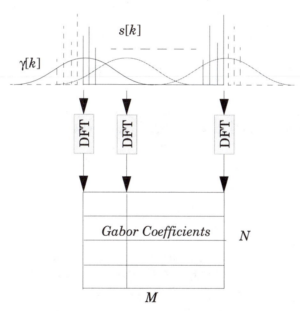

Fig. 3–6 Periodic Discrete Gabor transformation

The terms ΔM and ΔN in (3.20) denote discrete time and frequency sampling steps, respectively. The oversampling rate is defined by

$$a = \frac{L}{\Delta M \Delta N} \tag{3.22}$$

It is called critical sampling when $a = 1$ and oversampling when $a > 1$. For the stable reconstruction, the sampling rate must be greater or equal to one. In Wexler and Raz's original paper [185], it was required that $\Delta M M = \Delta N N = L^{*}$. In this case, M and N are equal to the number of sampling points in time and frequency domains, respectively. The product MN is equal to the total number of the Gabor coefficients. Rewriting (3.22) obtains

$$\text{oversampling rate } (\alpha) \; = \; \frac{\text{number of Gabor coefficients (MN)}}{\text{number of distinct samples (L)}}$$

which shows that the sampling rate is equal to the ratio between the total number of Gabor coefficients and the number of distinct samples. For critical sampling, the number of the Gabor coefficients is equal to the number of distinct samples. In the oversampling case, the number of the Gabor coefficients is greater than the number of samples. In this case, the resulting Gabor expansion is redundant.

If we restrict $\Delta N N = L$, (3.20) and (3.21) can be rewritten as

$$\tilde{s}[k] \; = \; \sum_{m=0}^{M-1} \sum_{n=0}^{N-1} \tilde{C}_{m,n} \, \tilde{h}[k - m\Delta M] W_N^{nk} \tag{3.23}$$

and

$$\tilde{C}_{m,n} \; = \; \sum_{k=0}^{L-1} \tilde{s}[k] \tilde{\gamma}^{*}[k - m\Delta M] W_N^{-nk} \tag{3.24}$$

where N can be considered the number of frequency bins or number of frequency channels. Eq. (3.24) implies that the Gabor coefficients are periodic in n, e.g.,

$$\tilde{C}_{m,n} = \tilde{C}_{m,n+iN} \qquad \forall \, l \in Z \tag{3.25}$$

Substituting $L = \Delta N N$ into (3.22) yields

$$\alpha \; = \; \frac{N}{\Delta M} \geq 1 \tag{3.26}$$

which says that for the stable reconstruction, the time sampling step ΔM has to be smaller or equal to the number of frequency bins or frequency channels N.

---

[*] Li and Qian show [116] that the requirement $\Delta N N = L$ is not necessary. ΔN in fact can be any integer. Consequently, we could have more freedom in choosing the oversampling rate. On the other hand, however, the implementation may become complicated and less efficient. For example, we may no longer be able to use the efficient FFT algorithm to compute the Gabor expansion as well as the sampled short-time Fourier transform.

The remaining question is how to compute $\gamma[k]$ for a given $h[k]$ and sampling steps. Substituting (3.24) into the right side of (3.23) yields

$$\tilde{s}[k] = \sum_{k'=0}^{L-1} \tilde{s}[k'] \sum_{m=0}^{M-1} \sum_{n=0}^{N-1} \tilde{\gamma}^*[k'-m\Delta M]\tilde{h}[k-m\Delta M]W_N^{n(k-k')} \tag{3.27}$$

Hence, the periodic discrete Gabor expansion exists if and only if the double summation is equal to the delta function, i.e.,

$$\sum_{m=0}^{M-1} \sum_{n=0}^{N-1} \tilde{\gamma}^*[k'-m\Delta M]\tilde{h}[k-m\Delta M]W_N^{n(k-k')} = \delta[k-k'] \tag{3.28}$$

The double summation form is not very pleasant to work with. Eq. (3.28) does not provide a clue to solving $\gamma[k]$. By the discrete Poisson-sum formula, Eq. (3.28) can be reduced to a single summation [185], such as

$$\sum_{k=0}^{L-1} \tilde{h}[k+qN]W_{\Delta M}^{-pk}\tilde{\gamma}^*[k] = \frac{\Delta M}{N}\delta[p]\delta[q] \tag{3.29}$$

where $0 \le p < \Delta M$ and $0 \le q < \Delta N$. Eq. (3.29) usually is considered the discrete version of *the Wexler-Raz identity.*

PROOF

Expanding the left side of (3.28) obtains

$$\sum_{m=0}^{M-1} \tilde{\gamma}^*[k'-m\Delta M]\tilde{h}[k-m\Delta M]N\sum_{q=0}^{\Delta N-1}\delta[k-k'-qN]$$

$$= N\sum_{q=0}^{\Delta N-1}\delta[k-k'-qN]\sum_{m=0}^{M-1}\tilde{\gamma}^*[k'-m\Delta M]\tilde{h}[k'+qN-m\Delta M] \tag{3.30}$$

Applying the discrete Poisson-sum formula (2.76) to the second summation,

$$\sum_{m=0}^{M-1}\tilde{\gamma}^*[k-m\Delta M]\tilde{h}[k+qN-m\Delta M]$$

$$= \frac{1}{\Delta M}\sum_{k=0}^{\Delta M-1}\left\{\sum_{i=0}^{L-1}\tilde{h}[k+qN]W_{\Delta M}^{-pk}\tilde{\gamma}^*[k]\right\}W_{\Delta M}^{nk}$$

which implies that (3.28) holds if and only if

$$\sum_{i=0}^{L-1}\tilde{h}[k+qN]W_{\Delta M}^{-pk}\tilde{\gamma}^*[k] = \frac{\Delta M}{N}\delta[p]\delta[q] \qquad for \ \ 0\le p<\Delta M \qquad 0\le q<\Delta N$$

Eq. (3.29) can also be formulated in terms of matrix computation, i.e.,

$$H\vec{\gamma}^* = \vec{\mu} \tag{3.31}$$

where H is a $\Delta M \Delta N$-by-L matrix, whose entries are defined as

$$h_{p\Delta M + q, k} \equiv \tilde{h}[k + qN]W_{\Delta M}^{-pk} \tag{3.32}$$

$\vec{\mu}$ is the $\Delta M \Delta N$ dimensional vector given by

$$\vec{\mu} = (\underbrace{\frac{\Delta M}{N}, 0, 0 \ldots 0}_{\Delta M \Delta N})^T$$

Eq. (3.31) shows that the dual function $\gamma[k]$ is no more than the solution of a linear system. The necessary and sufficient condition of the existence of the solution of (3.31) is that $\vec{\mu}$ is in the range of H. At the critical sampling case, $\Delta M \Delta N = L$, H is an L-by-L square matrix. Thereby, the solution is unique if it exists. For oversampling, $\Delta M \Delta N < L$, (3.31) is an underdetermined system and thereby the solution is not unique in general. In any case, (3.31) provides a feasible way to find the discrete dual function $\gamma[k]$.

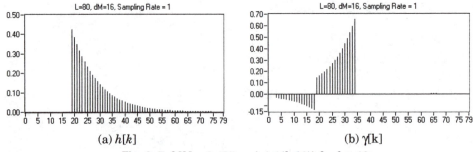

(a) $h[k]$ (b) $\gamma[k]$

Fig. 3–7 $h[k] = 0.425\exp\{-0.1(k-19)\}$ for $k \geq 19$

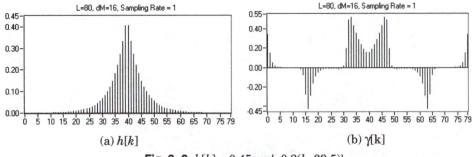

(a) $h[k]$ (b) $\gamma[k]$

Fig. 3–8 $h[k] = 0.45\exp\{-0.2(k-39.5)\}$

Fig. 3–7 and Fig. 3–8 illustrate the one- and two-sided normalized exponential sequences and their corresponding biorthogonal functions. Based on the algorithm discussed in this section, we virtually can compute dual functions for an arbitrarily given Gabor elementary function.

3.4 Discrete Gabor Expansion

In the preceding section, we introduced the periodic discrete Gabor expansion that requires the signal, analysis function, and synthesis function to have the same length. This is inconvenient (even impractical) in many applications, particularly when the number of samples L is large. For L-point samples, (3.31) essentially implies an L-by-L linear system. Therefore, it is desirable that lengths of analysis and synthesis function are independent of the length of samples so that we can use short windows to process arbitrarily long data. When this is achieved, the discrete Gabor expansion can then be used in many more signal process applications where typical signals are long.

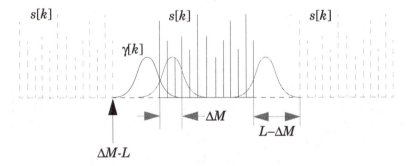

Fig. 3–9 Due to zero padding, the Gabor transform can be computed without rolling over either signal $s[k]$ or analysis function $\gamma[k]$. The auxiliary periodic sequences actually release the periodic constraint.

Assume the length of the signal $s[k]$ is L_s and the lengths of Gabor elementary function $h[k]$ and dual function $\gamma[k]$ are L. Let's build auxiliary periodic sequences as

$$\hat{s}[k] = \hat{s}[k + iL_0] = \begin{cases} 0 & -(L - \Delta M) \le k < 0 \\ s[k] & 0 \le k < L_s \end{cases} \tag{3.33}$$

$$\hat{h}[k] = \hat{h}[k + iL_0] = \begin{cases} h[k] & 0 \le k < L \\ 0 & L \le k < (L_s - \Delta M) \end{cases} \tag{3.34}$$

$$\hat{\gamma}[k] = \hat{\gamma}[k + iL_0] = \begin{cases} \gamma[k] & 0 \le k < L \\ 0 & L \le k < (L_s - \Delta M) \end{cases} \tag{3.35}$$

where all have the same period $L_0 = L_s + L - \Delta M$. The periodic discrete Gabor transform can then be plotted as Fig. 3–9.

Because of zero padding, we can compute the Gabor transform without rolling over either the signal $s[k]$ or analysis function $\gamma[k]$. The auxiliary periodic sequences defined in (3.33), (3.34), and (3.35) actually release the periodic constraint. Substituting auxiliary periodic sequences into the form of the periodic discrete Gabor expansion (3.23) and (3.24) yield

$$\tilde{C}_{m,n} = \sum_{k=\Delta M - L}^{L_s - 1} \hat{s}[k]\hat{\gamma}^*[k - m\Delta M]W_N^{-nk} \tag{3.36}$$

and

$$\hat{s}[k] = \sum_{m=m_0}^{M-1} \sum_{n=0}^{N-1} \tilde{C}_{m,n}\hat{h}[k - m\Delta M]W_N^{nk} \tag{3.37}$$

where

$$m_0 = \frac{\Delta M - L}{\Delta M} \tag{3.38}$$

The oversampling rate is $a = N/\Delta M$. For the perfect reconstruction, the time sampling step ΔM has to be less or equal to the number of frequency channels N. The total number of time sampling points is the smallest integer that is larger than or equal to $L_0/\Delta M$. Because of zero padding, the oversampling rate is equal to the ratio between the number of the Gabor coefficients and the number of zero padded samples $s[k]$.

With L remaining finite, letting $L_s \to \infty$ and thereby $L_0 \to \infty$, (3.36) and (3.37) directly lead to the Gabor expansion pair for discrete-time infinite sequence, i.e.,

$$\tilde{C}_{m,n} = \sum_{k=0}^{\infty} s[k]\gamma^*[k - m\Delta M]W_N^{-nk} \tag{3.39}$$

and

$$s[k] = \sum_{m=m_0}^{\infty} \sum_{n=0}^{N-1} \tilde{C}_{m,n}h[k - m\Delta M]W_N^{nk} \tag{3.40}$$

The remaining question is how to compute the dual function $\gamma[k]$. Substituting (3.34), and (3.35) into (3.29) and writing the resulting formula in the form of matrices, then we have

$$H\vec{\gamma_0}^* = \vec{\mu_0} \tag{3.41}$$

where $\vec{\gamma_0}$ is an L_0-by-1 vector and

$$\vec{\mu_0} = \left(\frac{\Delta M}{N}, 0, 0, \ldots\right)^T \tag{3.42}$$

H is a $\Delta M L_0 / N$-by-L_0 matrix that can be written as

$$\begin{vmatrix}
H_0 & H_1 & \ldots & H_{\Delta N-1} & 0 & \ldots & 0 & 0 \\
H_1 & \ldots & H_{\Delta N-1} & 0 & 0 & \ldots & 0 & H_0 \\
\ldots & H_{\Delta N-1} & 0 & 0 & \ldots & \ldots & H_0 & H_1 \\
H_{\Delta N-1} & 0 & 0 & 0 & 0 & \ldots & H_1 & \ldots \\
\ldots & & \ldots & & \ldots & & \ldots & \\
0 & 0 & 0 & H_0 & H_1 & \ldots & H_{\Delta N-1} & 0 \\
0 & 0 & H_0 & H_1 & \ldots & \ldots & 0 & 0 \\
0 & H_0 & H_1 & \ldots & H_{\Delta N-1} & \ldots & 0 & 0
\end{vmatrix} \tag{3.43}$$

where H_i are ΔM-by-N block matrices whose entries $h_{p,k}(i)$ are

$$h_{p,k}(i) = h_{p,k}(q+l) \equiv \hat{h}[(q+l)N + k]W_{\Delta M}^{-pk} \tag{3.44}$$

Because $\Delta N N = L$, $h_{p,k}(i) = 0$ for $i = (p+l) \geq \Delta N$. That is, $H_i = 0$ for $i \geq \Delta N$. In order to have dual functions $\gamma[k]$ and $h[k]$ have the same time support, we force the last $L_s - \Delta M$ elements of the vector $\vec{\gamma_0}$ to be zero, that is,

$$\vec{\gamma}_0 = \begin{vmatrix} \vec{\gamma} \\ \vec{0} \end{vmatrix} \qquad \text{where } \vec{0} = \underbrace{(0,0,0\ldots0)}_{L_s - \Delta M}^T \tag{3.45}$$

γ is an L-dimensional vector. Replacing $\vec{\gamma_0}$ in (3.41) by (3.45), (3.41) reduces to

$$
\begin{vmatrix}
H_0 & H_1 & \cdots & H_{\Delta N-1} \\
H_1 & \cdots & H_{\Delta N-1} & 0 \\
\cdots & H_{\Delta N-1} & 0 & \cdots \\
H_{\Delta N-1} & 0 & \cdots & 0 \\
0 & \cdots & 0 & H_0 \\
\cdots & & \cdots & \\
0 & H_0 & \cdots & H_{\Delta N-2}
\end{vmatrix}
\vec{\gamma}^* = \overline{H}\vec{\gamma}^* = \vec{\mu}
\tag{3.46}
$$

The matrix can be remembered by an auxiliary periodic sequence given by

$$
\overline{h}[k] = \overline{h}[k + l(2L - N)] \equiv \begin{cases} h[k] & 0 \le k < L \\ 0 & L \le k < 2L - N \end{cases}
\tag{3.47}
$$

Then, the entries of $\overline{H}$ can be defined in the same manner as in the case of the periodic discrete Gabor expansion (3.32), i.e.,

$$
\overline{h}_{p\Delta M + q, k} \equiv \overline{h}[k + qN]W_{\Delta M}^{-pk}
\tag{3.48}
$$

Consequently, (3.46) can be written as

$$
\sum_{k=0}^{L-1} \overline{h}[k + qN]W_{\Delta M}^{-pk}\gamma^*[k] = \frac{\Delta M}{N}\delta[p]\delta[q]
\tag{3.49}
$$

where $0 \le p < \Delta M$ and $0 \le q < 2(\Delta N-1)$. The significance of Eq. (3.49) is that it is independent of the signal length. It guarantees that the dual functions, $h[k]$ and $\gamma[k]$, have the same time support.

$\overline{H}$ in (3.46) is a K-by-L matrix, where

$$
K = \Delta M(2\Delta N - 1) = 2L\frac{\Delta M}{N} - \Delta M = \frac{2L}{a} - \Delta M
\tag{3.50}
$$

where a denotes the oversampling rate. Therefore, (3.46) is an underdetermined system when

$$
\frac{2L}{a} - \Delta M < L
\tag{3.51}
$$

that is, a is larger than $2L/(L+\Delta M)$.

It is interesting to note that the periodic discrete Gabor expansion and discrete Gabor expansion have a similar formula of computing the dual functions (see (3.31) and (3.46)) except for the structures of functions H. In the case of the periodic Gabor expansion, H is made up of the periodic window function $h[k]$

directly. For the discrete Gabor expansion, H is constituted by the periodic auxiliary function $\bar{h}[k]$ which is the zero padded window function $h[k]$.

Finally, $\gamma[k]$ derived for the discrete Gabor expansion is a subset of $\vec{\gamma}_0$ in (3.45), which is a special solution of the periodic discrete Gabor expansion introduced in Section 3.3. Because we force the last L_s–ΔM elements of the vector $\vec{\gamma}_0$ to be zero, the existence of $\gamma[k]$ is much more restricted than that in periodic cases, in particular, for the critical sampling $\Delta M = N$. Because the critical sampling does not introduce any redundancy, it plays an important role in many signal processing applications, such as the maximally decimated linear systems. Then, the important issue is the existence of the critically sampled discrete Gabor expansion. In what follows, we only give the results without derivations. The reader can find the rigorous mathematical treatment in Appendix A.

For clarity of presentation, let's define the operation $\otimes$ by

$$e_0 \otimes e_1 \otimes e_2 \otimes \ldots e_m = \left\{ \begin{array}{ll} non-zero & \text{iff one term is not zero} \\ 0 & \text{otherwise} \end{array} \right. \tag{3.52}$$

If a set of numbers $\{e_0, e_1, e_2, \ldots, e_m\}$ satisfies the condition

$$e_0 \otimes e_1 \otimes e_2 \otimes \ldots \otimes e_m \neq 0 \tag{3.53}$$

then we call this set of numbers *exclusively non-zero*.

Now, we state that *for critical sampling, the biorthogonal function $\gamma[k]$ of the discrete Gabor expansion exists iff*

$$h[k] \otimes h[N+k] \otimes h[2N+k] \otimes \ldots \otimes h[(\Delta N - 1)N + k] \neq 0 \tag{3.54}$$

where $0 \leq k < N$ and $\Delta N N = L$. If $h[k]$ satisfies Eq. (3.54), then $\gamma[k]$ is uniquely determined by

$$\gamma[mN + k] = \left\{ \begin{array}{ll} \dfrac{1}{Nh[mN+k]} & h[mN+k] \neq 0 \\ 0 & \text{otherwise} \end{array} \right. \tag{3.55}$$

where $0 \leq m < \Delta N$ and $0 \leq k < N$.

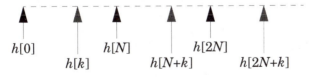

Fig. 3–10 The locations of $h[mN+k]$ at the critical sampling, where $0 \leq k < N$ and $0 \leq m < \Delta N$.

Fig. 3–10 depicts the locations of $h[mN+k]$. Eq. (3.54) implies that $h[mN+k]$ can only contain N non-zeros. Moreover, the non-zero point can only be one k for all different m. When $0 \le m < \Delta N = 1$, that is, $N = L$ (the number of frequency channels is equal to the length of the function $h[k]$), then the necessary and sufficient condition of the existing of dual function is simply that

$$h[k] \ne 0 \qquad \forall \, k \in [0, L)$$

which is illustrated in Fig. 3–11. In fact, this is exactly the case of non-overlap windowed Fourier transform.

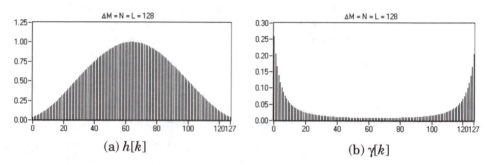

(a) $h[k]$ (b) $\gamma[k]$

Fig. 3–11 Biorthogonal sequences at the critical sampling ($\Delta M = N = L$).

Fig. 3–12 plots $h[k]$ and $\gamma[k]$ for $\Delta M = N = L/2 = 64$. In this case, non-zero points are $h[32]$ to $h[63]$ and $h[64]$ to $h[64+31]$, which obviously satisfies the conditions described by (3.54) and (3.55).

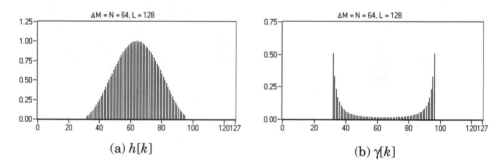

(a) $h[k]$ (b) $\gamma[k]$

Fig. 3–12 Biorthogonal sequences at the critical sampling ($\Delta M = N = L/2$).

3.5 Orthogonal-Like Gabor Expansion

As shown earlier, given $h[k]$ and the sampling steps, the solution of $\gamma[k]$ in general is not unique. Then, the question is how to select the $\gamma[k]$ that best meets our goal. Recall that the Gabor coefficients $C_{m,n}$ are the sampled short-time Fourier transform with the window function $\gamma[k]$. This means that the window function $\gamma[k]$ has to be localized in the joint time-frequency domain. Otherwise, the Gabor coefficients $C_{m,n}$, inner product of s[k] and $\gamma[k]$, would not characterize the signal's local behavior.

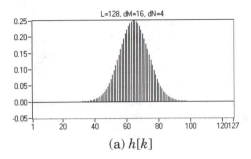

(a) $h[k]$

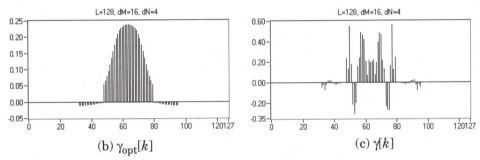

(b) $\gamma_{opt}[k]$ (c) $\gamma[k]$

Fig. 3–13 Parts (b) and (c) are dual functions of $h[k]$ in (a). They both yield the perfect reconstruction. $\gamma_{opt}[k]$ in (b) is optimally close to Gaussian-type function $h[k]$ in (a)[*].

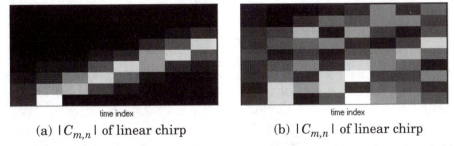

(a) $|C_{m,n}|$ of linear chirp   (b) $|C_{m,n}|$ of linear chirp

Fig. 3–14 (a) is computed by $\gamma_{opt}[k]$, which well presents the linear chirp signal. (b) is computed by $\gamma[k]$ plotted in Fig. 3–13c, which does not provide desired information of the linear chirp signal in the joint time-frequency domain. However, both Gabor coefficients will lead to the perfect reconstruction by using the same synthesis function $h[k]$.

Second, the behaviors of $\gamma[k]$ and $h[k]$, such as time/frequency centers and time/frequency resolution, have to be close. Suppose that $h[k]$ and its Fourier

[*] The algorithm of computing different dual functions for a same given function $h[k]$ is introduced in Appendix B.

transform are centered in $k = 0$ and $\theta = 0$. If $\gamma[k]$ and $h[k]$ are significantly different, for instance, they have completely different time or frequency centers, then
$C_{m,n}$ will not reflect the signal behaviors in the vicinity of $(mT, n\Omega)$.

Fig. 3–13 depicts two different dual functions $\gamma[k]$ that both correspond to
the same Gabor elementary function $h[k]$ in (a). The shape of $\gamma[k]$ in (b) is optimally close to that of $h[k]$, whereas the shape of $\gamma[k]$ in (c) significantly differs
from $h[k]$. Although both $\gamma[k]$ will lead to a perfect reconstruction with the same
$h[k]$, the resulting Gabor coefficients have substantial differences.

Fig. 3–14 illustrates the magnitude of the Gabor coefficients computed by
two different $\gamma[k]$ for the linear chirp signal. (a) is computed by the optimal $\gamma[k]$.
Because $h[k]$ in Fig. 3–13a is concentrated in the joint time and frequency
domain, the optimal $\gamma[k]$ is also concentrated. Consequently, the resulting
Gabor coefficients well present the monotone linear chirp signal. (b) is computed
by $\gamma[k]$ plotted in Fig. 3–13c, whose shape significantly differs from the Gabor
elementary function $h[k]$ in Fig. 3–13a. Due to the bad shape of the analysis
function, the resulting Gabor coefficients plotted in Fig. 3–14b do not properly
describe the time-varying nature of the linear chirp.

Because the Gabor elementary function $h[k]$ is the Gaussian function that
is optimally concentrated in the joint time-frequency domain, the natural selection of $\gamma[k]$ is that whose shape is closest to $h[k]$, in the sense of the least square
error[*], i.e.,

$$\Gamma = \min_{\vec{H\gamma} = \vec{\mu}} \sum_{k=0}^{L-1} \left| \frac{\gamma[k]}{\|\gamma\|} - h[k] \right|^2 \tag{3.56}$$

where

$$\|\gamma\| = \sqrt{\sum_{k=0}^{L-1} |\gamma[k]|^2} \tag{3.57}$$

and $h[k]$ is a normalized function, that is, $\|h[k]\|^2 = 1$. For small Γ, $\gamma[k] \approx ah[k]$
where a is a real-valued constant. In this case, the discrete Gabor expansion can
be written as

$$\tilde{C}_{m,n} \approx a \sum_{k=0}^{\infty} s[k] h^*[k - m\Delta M] W_N^{-nk} \tag{3.58}$$

---

[*] Because (3.31) and (3.46) have the same form, discussions in this section apply for both periodic and non-periodic cases.

and

$$s[k] = \sum_{m=m_0}^{\infty} \sum_{n=0}^{N-1} \tilde{C}_{m,n} h[k - m\Delta M] W_N^{nk} \tag{3.59}$$

Obviously, as long as $h[k]$ is localized in the joint time-frequency domain, the Gabor coefficients $C_{m,n}$ will well depict the signal's local time-frequency properties. Although the set of $\{h_{m,n}[k]\}$ is not orthogonal and redundant, the coefficients $C_{m,n}$ are still good approximations of signal orthogonal projection on $\{h_{m,n}[k]\}$. Therefore, (3.59) is called an *orthogonal-like Gabor expansion*. The remaining problem is how to solve (3.56).

Expanding Eq. (3.56) obtains [118]

$$\Gamma = \min_{\overrightarrow{H\gamma} = \overrightarrow{\mu}} \sum_{k=0}^{L-1} \left\{ \frac{\gamma^2[k]}{\|\gamma\|} - 2\frac{h[k]\gamma[k]}{\|\gamma\|} + h^2[k] \right\} \tag{3.60}$$

$$= \min_{\overrightarrow{H\gamma} = \overrightarrow{\mu}} \left\{ 2 - \frac{2}{\|\gamma\|} \sum_{k=0}^{L-1} h[k]\gamma[k] \right\}$$

Because (see (3.49))

$$\sum_{k=0}^{L-1} \bar{h}[k + qN] W_{\Delta M}^{-pk} \tilde{\gamma}^*[k] = \frac{\Delta M}{N} \delta[p]\delta[q]$$

(3.60) reduces to

$$\Gamma = \min_{\overrightarrow{H\gamma} = \overrightarrow{\mu}} \left(1 - \frac{1}{\|\gamma\|} \frac{\Delta M}{N} \right) \tag{3.61}$$

which implies that the solution is the minimum energy of $\gamma[k]$. According to the matrix analysis theory, $\gamma[k]$ exists as long as vector $\overrightarrow{\mu}$ is in the range of matrix H. If matrix H is of full-row rank, the minimum energy of $\gamma[k]$ is equal to the pseudo inverse of H, i.e.,

$$\overrightarrow{\gamma}^*_{\text{opt}} = H^T(HH^T)^{-1}\overrightarrow{\mu} \tag{3.62}$$

When the rank of H is less than the number of rows, we can employ the singular value decomposition (SVD) to alleviate the rank deficiency problem.

Assume that the rank of H is p and the number of rows is q, where $q \geq p$. Applying SVD to matrix H, (3.46) becomes

$$USV\overrightarrow{\gamma}^* = \overrightarrow{\mu} \tag{3.63}$$

where matrices $U \in C^{q,q}$ and $V \in C^{L,L}$ are unitary. $S \in R^{q,L}$. Multiplying U^T to both sides of (3.63) yields

$$\left| \begin{matrix} S^0_{p \times p} & 0_{p \times (l-p)} \\ 0_{(q-p) \times p} & 0_{(q-p) \times (l-p)} \end{matrix} \right| \left\| \begin{matrix} V^0_{p \times L} \\ V^1_{(L-p) \times L} \end{matrix} \right\| \vec{\gamma}^* = \left| U^0_{q \times p} \quad U^1_{q \times (q-p)} \right|^T \vec{\mu} \qquad (3.64)$$

where $S^0 \in R^{p,p}$ is a diagonal matrix with non-negative main diagonal entries. The solution of (3.64) exists when $\vec{\mu}$ is orthogonal to U^1 (consistency condition). When the consistency condition is satisfied, we rewrite (3.64) as

$$\hat{H} \vec{\gamma}^* = \hat{\mu} \qquad (3.65)$$

where $\hat{H} = S^0 V^0 \in C^{P,L}$ has full row rank p and $\hat{\mu} = (U^0)^T \vec{\mu}$. Then, the minimum energy solution of $\gamma[k]$ is

$$\vec{\gamma}^*_{opt} = \hat{H}^T (\hat{H} \hat{H}^T)^{-1} \hat{\mu} \qquad (3.66)$$

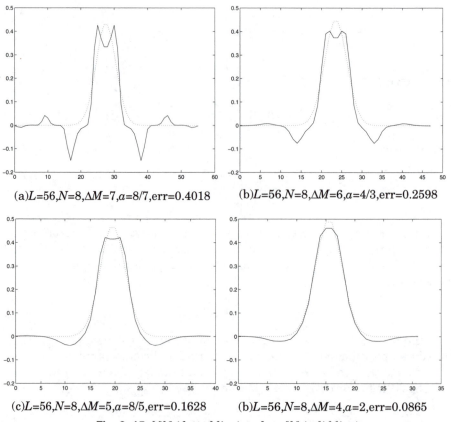

(a)$L=56, N=8, \Delta M=7, a=8/7$,err=0.4018   (b)$L=56, N=8, \Delta M=6, a=4/3$,err=0.2598

(c)$L=56, N=8, \Delta M=5, a=8/5$,err=0.1628   (b)$L=56, N=8, \Delta M=4, a=2$,err=0.0865

Fig. 3–15 $h[k]$ (dotted line) and $\gamma_{opt}[k]$ (solid line).

Fig. 3–15 depicts the Gaussian function and corresponding dual functions at different sampling schemes. In order to utilize the FFT (fast Fourier transform), the number of frequency bins (or channels) N is usually chosen as two's power. For instance, $N = 2^k$, $k = 1, 2, 3, \ldots$ The oversampling rate is determined by $a = N/\Delta M$, where ΔM denotes the time-sampling step. The length of $h[k]$ and $\gamma_{opt}[k]$, L, has to be divided by both N and ΔM. That is, L/N and $L/\Delta M$ have to be integers.

In general, the difference between $h[k]$ and $\gamma_{opt}[k]$ decreases as the oversampling rate increases. To faithfully characterize a signal's local properties in the joint time-frequency domain, it is critical to make $\gamma_{opt}[k]$ and $h[k]$ as close as possible. The minimum difference between $\gamma_{opt}[k]$ and $h[k]$, Γ in (3.56), is related to the selection of $h[k]$ and sampling steps. If the Gaussian function is used, for instance,

$$h[k] = \left(\frac{\alpha}{\pi}\right)^{\frac{1}{4}} \exp\left\{-\frac{\alpha}{2}k^2\right\} \tag{3.67}$$

the minimum difference between $\gamma_{opt}[k]$ and $h[k]$ is observed when

$$\alpha = \frac{2\pi}{\Delta M N} \tag{3.68}$$

The discrete Gabor expansion and the concept of orthogonal-like Gabor expansion were first proposed by Qian et al. (see [148] and [149]) In addition to seeking $\gamma[k]$ that is optimally close to $h[k]$, the algorithm introduced in this section could be easily modified to have $\gamma[k]$ optimally close to an arbitrary function, i.e.,

$$\Gamma = \min_{\vec{H\gamma} = \vec{\mu}} \sum_{k=0}^{L-1} \left|\frac{\gamma[k]}{\|\gamma\|} - d[k]\right|^2 \tag{3.69}$$

where $d[k]$ is a normalized target function. Such generalized optimization is very useful, in particular, for those applications where analysis and synthesis functions are desired to have different properties. For instance, we could use the Gabor expansion for filter bank design. In many filter bank applications, analysis and synthesis functions usually have different requirements. In this case, while one filter can be predetermined by $h[k]$, the other could be solved via (3.69). The general solution of (3.69) is discussed in Appendix B at the end of this chapter.

3.6 Fast Algorithm of Computing Dual Functions

In Sections 3.3 and 3.4, we discuss the algorithms of computing the dual function $\gamma[k]$ for the periodic discrete Gabor expansion and infinite discrete Gabor expansion, respectively. In both cases, $\gamma[k]$ are the solutions of the linear complex systems. In this section, we shall show that those complex matrices in fact could be converted into real-valued matrices and thereby save a considerable amount of computations [118]. The following discussions are mainly based on the case of the infinite samples (3.49), but the results should be easily extended to the periodic cases (3.29) simply by replacing the auxiliary function $\bar{h}[k]$ by the periodic function $\tilde{h}[k]$.

Given the finite Gabor elementary function $h[k]$, the dual function $\gamma[k]$ for the discrete Gabor expansion can be computed via Eq. (3.49)

$$\sum_{k=0}^{L-1} \bar{h}[k+qN]W_{\Delta M}^{-pk}\gamma^*[k] = \frac{\Delta M}{N}\delta[p]\delta[q]$$

where $0 \le p < \Delta M$ and $0 \le q < 2(\Delta N-1)$. The auxiliary periodic function $\bar{h}[k]$ is defined in (3.47). Therefore, computing $\gamma[k]$ is to solve the linear complex system with $\Delta M(2\Delta N-1)$-by-L complex matrix H.

Let's take the inverse DFT with respect to (3.49), i.e.,

$$\sum_{k=0}^{L-1} \bar{h}[k+qN]\gamma^*[k]\frac{1}{\Delta M}\sum_{p=0}^{\Delta M-1} W_{\Delta M}^{-p(k-k')} = \frac{\Delta M}{N}\delta[q]\frac{1}{\Delta M}\sum_{p=0}^{\Delta M-1}\delta[p]W_{\Delta M}^{-pk'} \qquad (3.70)$$

Because

$$\frac{1}{\Delta M}\sum_{p=0}^{\Delta M-1} W_{\Delta M}^{-p(k-k')} = \delta(k-k'-l\Delta M) \qquad (3.71)$$

Eq. (3.70) reduces to

$$\sum_{k=0}^{L-1} \bar{h}[k+p\Delta M+qN]\gamma^*[k+p\Delta M] = \frac{1}{N}\delta[q] \qquad (3.72)$$

where $0 \le k < \Delta M$ and $0 \le q < 2\Delta N-1$. Eq. (3.72) suggests that $\gamma[k]$ can be computed by ΔM-separated real-valued linear systems, i.e.,

$$H_k\vec{\gamma}^*_k = \vec{\mu}_k \qquad (3.73)$$

where H_k are $(2\Delta N-1)$-by-M matrices with the entries:

$$h_{q,p}(k) \equiv \bar{h}[k+p\Delta M+qN] \qquad (3.74)$$

$\vec{\gamma}_k$ is an M-dimensional vector with the entries:

$$\gamma_p(k) = \gamma[k+p\Delta M] \qquad (3.75)$$

and

$$\bar{\mu}_k = (\underbrace{N^{-1},0,0...0}_{2\Delta N-1})^T$$

While (3.49) is a $\Delta M(2\Delta N-1)$-by-L linear *complex* system, the solution of (3.72) is a ΔM independent $(2\Delta N-1)$-by-M linear *real* system. Therefore, considerable computations are saved when (3.72) is used. Moreover, it is interesting to note that the minimum norm of (3.31) is equal to the minimum norms computed by each individual linear system in (3.73). We leave the proof for the reader to exercise.

Summary

In this chapter, we discussed the short-time Fourier transform (STFT) and Gabor expansion. Although Gabor did not investigate the inverse of the STFT, the Gabor expansion turns out to be the most elegant algorithm of computing the inverse of a sampled STFT. The Gabor expansion has been thoroughly studied in the mathematics literature for a long time. Most of the analysis was confined to the continuous-time cases, which has a wider scope for deeper mathematical issues. The discrete Gabor expansion, on the other hand, is relatively easier to understand. It could be implemented with the help of elementary linear algebra. In Sections 3.3 and 3.4, we introduced Gabor expansions for periodic and infinite discrete-time samples, respectively. Although the structure of H matrices in (3.31) and (3.46) are different, the dual functions in both cases could be solved by similar linear systems. In Section 3.5, we bring up the concept of orthogonal-like, which not only is fundamental for time-frequency analysis, but also is important for time-varying filtering. We shall elaborate on this subject in Chapter 9.

Wavelets[*]

$\boldsymbol{T}$he most common technique of studying some properties of signals that are not obvious in the time domain is to compare the given time function with a set of prudently selected elementary functions. For example, to explore a signal's periodic property, traditionally we compare the time signal $s(t)$ with a set of harmonically related complex sinusoidal functions $\exp\{j2\pi nt/T\}$. Because each individual complex harmonic sinusoidal function $\exp\{j2\pi nt/T\}$ corresponds to a particular frequency $2\pi n/T$, the Fourier transform, the measure of similarities between $s(t)$ and $\exp\{j2\pi nt/T\}$, reflects the signal's behavior at frequency $2\pi n/T$. By changing the parameter n, we can obtain all different frequency tick marks. In addition to complex harmonic sinusoidal functions, we can also build frequency tick marks by scaling the time index t of a given elementary function $\psi(t)$, as introduced in Chapter 2. Scaling signal in the time domain results in inversely scaling in the frequency domain. When the dilated (scaled) and translated (time-shifted) elementary functions $\psi(a^{-1}(t-b))$ are employed to measure the given signal, the resulting presentation is named *time-scale representations* or *wavelets*. The elementary function $\psi(t)$ is known as the *mother wavelet*.

Although the original idea can be traced back to the Haar transform at the beginning of the century, wavelets were not popular until the early eighties. About ten years ago, researchers from geophysics, theoretical physics, and mathematics developed a solid foundation for "wavelets" (see [61] and [125]). Since then, this topic has been treated in considerable detail by numerous researchers in both the mathematics and engineering literature. In particular, Mallat (see [120] and [121]) and Meyer [126] discovered a close relationship between wavelet

---

[*]This chapter is co-authored with Xiang-Gen Xia, Hughes Research Laboratories, Malibu, California 90265.

and mutiresolution analysis structure, which leads to a simple way of calculating the mother wavelet. Their work also established a connection between continuous-time wavelets and digital filter banks. Daubechies (see [34], [35], and [36]) developed a systematic technique for generating finite-duration orthonormal wavelets with FIR (Finite Impulse Response) filter banks. Those results have triggered tremendous interest in the mathematics as well as signal processing communities. Due to the efficiency in representing non-stationary signals, such as speech and image/video, and many other interesting properties, wavelets has become one of the most active research areas.

Although both wavelets and STFT can be used to study a signal's local behaviors, they have substantial differences in many aspects. Unlike the STFT and Gabor expansion, which can be realized with the help of a fair amount of elementary matrix theory, to apply wavelet analysis one needs a certain level of mathematical preparation. Because the aim of this book is to university students and practicing engineers/scientists, in what follows, we shall only introduce the most fundamental ideas without delving into mathematical details. Although our presentations are limited in the orthonormal wavelet transform, this chapter, we believe, would serve as a good introduction for those who are interested in this exciting field.

In Section 4.1, we introduce basic concepts of wavelets. In particular, we emphasize its differences from STFT. For STFT-based signal decomposition (e.g., the Gabor expansion), we can virtually use any function as the window function. For wavelet decomposition, however, the valid mother wavelet has to satisfy an admissibility condition. Therefore, one main issue for wavelet analysis is how to generate a desired mother wavelet. A popular algorithm that was first studied by Mallat [121] and Meyer [126] is based upon the multiresolution analysis structure. To facilitate the reader's understanding of multiresolution analysis, in Section 4.2, we review a simple example, piecewise constant approximation. Then, in Section 4.3, we discussed mathematical beauty of the multiresolution analysis. Finally, in Section 4.4, we introduce digital implementations. As a matter of fact, in most real applications, we can well estimate wavelet transform by filter banks without explicitly calculating the mother wavelet function $\psi(t)$.

4.1   Continuous Wavelet Transform

In Chapter 2, we showed that there are usually two ways of building frequency tick marks to test the signal in the frequency domain. One is to use *harmonically related complex sinusoidal functions*, such as Fourier transform as well as Gabor transform. The other is achieved by scaling the time variable t of a given elementary function $\psi(t)$. If the center frequency, or the mean frequency, of the function $\psi(t)$ is ω_0, then the center frequencies of its time-scaled version (or dilation) $\psi(t/a)$ will be the reciprocal of ω_0, for instance, $<\omega> = \omega_0/a$, as shown in Example 2–9. When the dilated and shifted function $\psi(a^{-1}(t-b))$ is used as the tick marks to measure a signal's local behavior, the resulting presentation is named *wavelet transform* (WT), or the *continuous-time wavelet transform* (CWT) if the signal under consideration is a function of continuous-time, i.e.,

$$\text{CWT}(a, b) = \frac{1}{\sqrt{|a|}} \int s(t) \psi^* \left(\frac{t-b}{a} \right) dt \qquad a \neq 0 \qquad (4.1)$$

where $\psi(t)$ denotes the *mother wavelet*. The parameter a represents the scale index that is the reciprocal of the frequency. The parameter b indicates the time shifting (or translation). Suppose that $\psi(t)$ is centered at time zero and frequency ω_0. Then, its dilation and translation $\psi(a^{-1}(t-b))$ is centered at time b and frequency ω_0/a, respectively. Consequently, the quantity of CWT(a,b), inner product of $s(t)$ and $\psi(a^{-1}(t-b))$, reflects the signal's behavior in the vicinity of $(b, \omega_0/a)$. Therefore, *CWT(a,b)* could also be thought of as a function of time and frequency by

$$\text{CWT}(a, b) \Big|_{a = \frac{\omega_0}{\omega}, b = t} = \text{TF} \left(\frac{\omega_0}{\omega}, t \right) \qquad (4.2)$$

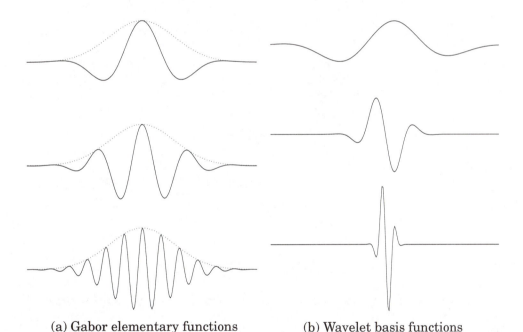

(a) Gabor elementary functions   (b) Wavelet basis functions

Fig. 4–1 In the Gabor transform (or STFT), all elementary functions have the same envelope, whereas the wavelet basis functions have different envelopes. The time and frequency resolutions of wavelet basis functions change with a scale factor.

In STFT, the "ruler" used to measure the signal's joint time-frequency property is made up of time-shift and frequency-modulated single prototype function $\gamma(\tau-t)\exp\{j\omega\tau\}$. Therefore, all elementary functions have the same envelope as shown in Fig. 4–1. Once $\gamma(t)$ is chosen, both the time and frequency resolutions of the elementary functions $\gamma(\tau-t)\exp\{j\omega\tau\}$ are fixed. On the other hand, the "ruler" used in the wavelet transform is obtained by the dilation and translation of a mother wavelet. Consequently, both time resolution and frequency resolutions of the basis function $\psi(a^{-1}(t-b))$ are functions of the scaling factor.

Let's use the standard deviation to characterize the signal's resolution. If time and frequency deviations of the mother wavelet $\psi(t)$ are Δ_t and Δ_ω, then the corresponding time and frequency deviations of $\psi(a^{-1}(t-b))$ are $a\Delta_t$ and Δ_ω/a, respectively (see Example 2–9). For the wavelet basis function $\psi(a^{-1}(t-b))$, the higher the time resolution (smaller a) is, the worse the frequency resolution. This suggests that the wavelet basis matches a long-duration signal with low oscillation or a short-duration signal with high oscillation. Fig. 4–2 illustrates STFT and WT tiling in the joint time-frequency domain. While the STFT tiling is linear, the WT tiling is logarithmic.

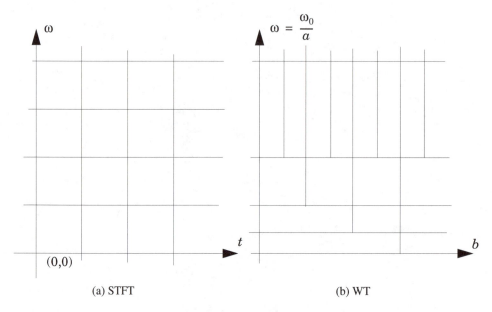

(a) STFT   (b) WT

Fig. 4–2 Tilings of STFT and WT.

It is interesting to note that for wavelet basis function $\psi(a^{-1}(t-b))$, the product of time resolution $a\Delta_t$ and frequency resolution Δ_ω/a is independent of the scale factor a. This indicates that the wavelet's function $\psi(a^{-1}(t-b))$ also obeys the uncertainty principle. Moreover, assume that the mean frequency of the mother wavelet $\Psi(t)$ is ω_0. Then, the mean frequency of $\psi(a^{-1}(t-b))$ is ω_0/a (see Example 2–9). Consequently, the ratio of frequency bandwidth $2\Delta_\omega/a$ (double standard deviation in frequency domain) and the mean frequency ω_0/a is unchanged with the scale a. That is,

$$Q = \frac{2\Delta_\omega/a}{\omega_0/a} = \frac{2\Delta_\omega}{\omega_0}$$

which naturally links the wavelet transform to the conventional constant Q analysis. Frequency responses of STFT and WV are plotted in Fig. 4–3 and Fig. 4–4, respectively.

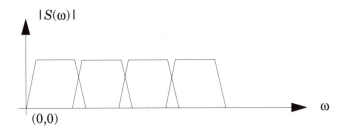

Fig. 4–3 Bandwidth of STFT is uniform in frequency domain.

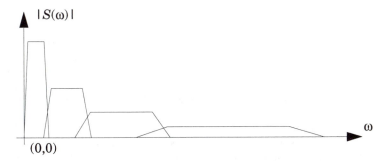

Fig. 4–4 Bandwidth of WV basis increases as frequencies increase.

In order to get a better feeling about the difference between STFT and WT, let's look at a simple example. Suppose that we have a signal that contains two time pulses and two frequency pulses, such as

$$s(t) = \delta(t - t_1) + \delta(t - t_2) + e^{j\omega_1 t} + e^{j\omega_2 t} \qquad (4.3)$$

Then its frequency presentation is

$$S(\omega) = e^{j\omega t_1} + e^{j\omega t_2} + 2\pi\delta(\omega - \omega_1) + 2\pi\delta(\omega - \omega_2)$$

Fig. 4–5 compares STFT and WT for the signal in (4.3). While time and frequency resolutions of STFT are uniform in the entire time-frequency domain, they vary in WT. At high frequencies, we have better time resolution and bad frequency resolution. At low frequencies, we have better frequency resolution and bad time resolution. However, the ratio of bandwidth and center frequency is constant.

The square of WT is commonly named *scalogram*, i.e.,

$$\text{SCAL}(a, b) = |\text{CWT}(a, b)|^2 \tag{4.4}$$

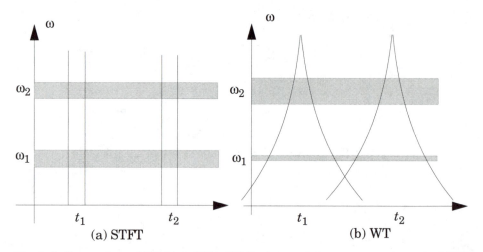

Fig. 4–5 Comparison of STFT and WT. (While time and frequency resolutions of STFT are uniform in the entire time-frequency domain, they vary in WT. Frequency/time resolution gets better at low/high frequencies and becomes worse at high/low frequencies.)

It is worthwhile to note that if we only want to analyze the signal and do not want to recover the original signal based upon the transformations, then the mother wavelet $\psi(t)$ in (4.1) could be any functions we like. When the perfect reconstruction is needed, the selection of the mother wavelet $\psi(t)$ will be much more restricted; it has to satisfy the *admissibility condition* given by

$$C_\psi = \frac{1}{2\pi}\int\frac{|\Psi(\omega)|^2}{|\omega|}d\omega < \infty \tag{4.5}$$

$\Psi(\omega)$ is the Fourier transform of the mother wavelet $\psi(t)$. Condition (4.5) implies $\Psi(0) = 0$. In other words, the mother wavelet $\psi(t)$ is of *bandpass*. Once $\psi(t)$ meets the admissibility condition, then the original signal $s(t)$ can be reconstructed by

$$s(t) = \frac{1}{C_\psi}\iint\frac{1}{a^2}\text{CWT}(a, b)\hat{\psi}\left(\frac{t-b}{a}\right)da\,db \tag{4.6}$$

which is named the inverse wavelet transform. $\hat{\psi}(t)$ is a dual function of $\psi(t)$. Usually, $\hat{\psi}(t) = \psi(t)$. Like the case of the Gabor expansion, the representation of continuous-time wavelet is redundant. The original signal could be completely reconstructed by the sampled CWT. The resulting presentation is known as the *wavelet series*. We shall discuss this issue in more detail in following sections.

In addition to the properties mentioned earlier, another important feature of wavelet transform is

$$\int |s(t)|^2 dt = \frac{1}{C_\psi} \int\int a^{-2} |\mathrm{CWT}(a, b)|^2 da\,db \tag{4.7}$$

which says that the weighted energy of the wavelet transform in the joint time-frequency plane is equal to the energy of the signal in the time domain. Traditionally, (4.7) is considered the counterpart of the Parseval's formula for the Fourier transform. It is worthwhile to note that such a relation in general does not hold for STFT unless the function $\gamma(t)$ is badly localized [36].

Finally, because the elementary functions used in STFT are complex harmonic sinusoidal functions, STFTs usually are complex. On the other hand, WT are real-valued as long as both signal $s(t)$ and mother wavelet $\psi(t)$ are real. This makes WT attractive for many applications.

4.2 Piecewise Approximation

The representation of the continuous-time wavelet is redundant. For a given arbitrary signal, in fact, we can completely characterize it by sampled CWT. Traditionally, we sample CWT in a dyadic grid, that is,

$$a = 2^{-m} \qquad b = n2^{-m} \tag{4.8}$$

Substituting (4.8) into (4.1) yields

$$d_{m,n} = \mathrm{CWD}(2^{-m}, n2^{-m}) = 2^{m/2} \int s(t)\psi^*(2^m t - n)dt = \int s(t)\psi^*_{m,n}(t)dt$$

where $\psi_{m,n}(t)$ are translated and dilated versions of the mother wavelet function $\psi(t)$ given by

$$\psi_{m,n}(t) = 2^{m/2}\psi(2^m t - n)$$

Based on the sampled wavelet transform $d_{m,n}$, we can reconstruct the original signal $s(t)$ by

$$s(t) = \sum_{m=-\infty}^{\infty} \sum_{n=-\infty}^{\infty} d_{m,n} \hat{\psi}_{m,n}(t) = \sum_{m=-\infty}^{\infty} \sum_{n=-\infty}^{\infty} \mathrm{CWD}(2^{-m}, n2^{-m})2^{m/2} \hat{\psi}(2^m t - n) \tag{4.9}$$

where $\hat{\psi}_{m,n}(t)$ are dual functions of $\psi_{m,n}(t)$.

One important issue in wavelet theory is how to determine $\psi_{m,n}(t)$ as well as $\hat{\psi}_{m,n}(t)$. For the sake of simplicity, let's limit our discussion to the case where

$\hat{\psi}_{m,n}(t) = \psi_{m,n}(t)$, which implies that $\{\psi_{m,n}(t)\}$ are orthonormal. In this case, (4.9) becomes

$$s(t) = \sum_{m=-\infty}^{\infty} \sum_{n=-\infty}^{\infty} d_{m,n}\,\psi_{m,n}(t) \tag{4.10}$$

which is commonly known as a *wavelet series*. The constants $d_{m,n}$ are wavelet series coefficients and computed by the regular inner products

$$d_{m,n} = \int s(t)\psi_{m,n}^*(t)dt \tag{4.11}$$

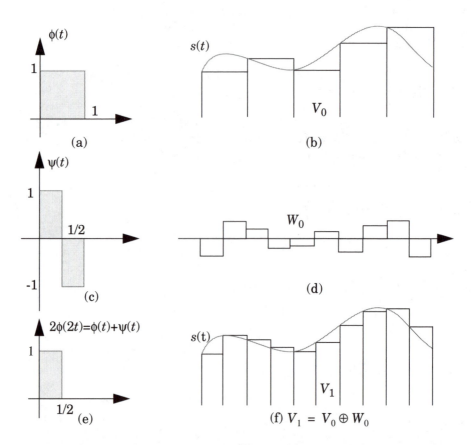

Fig. 4–6 Piecewise approximation: (a) elementary function of V_0, (b) V_0, (c) elementary function of W_0 ($\psi(t)$ that is orthogonal to $\phi(t)$), (d) W_0, (e) elementary function of V_1, (f) V_1 can be achieved by the coarse approximation V_0 plus the detail W_0.

One popular algorithm of computing the mother wavelet $\psi(t)$ is based on the concept of the multiresolution analysis, which was first studied by Mallat [121] and Meyer [126]. Before introducing abstract descriptions of multiresolution analysis, in what follows, let's first look at a simple example - piecewise approximation.

As shown in Fig. 4–6b and f, a given arbitrary signal $s(t)$ can be approximated at different levels by piecewise constant basis functions. For instance, the approximation depicted in (b) can be accomplished by translated piecewise constant function $\phi(t)$ in Fig. 4–6 a, where

$$\phi(t) = \begin{cases} 1 & 0 \leq t < 1 \\ 0 & \text{otherwise} \end{cases}$$

By the definition of $\phi(t)$, translated $\phi(t)$ are orthogonal to each other, i.e.,

$$\int \phi(t-n)\phi^*(t-n')dt = \delta(n-n') \qquad \text{for } n, n' \in Z$$

We define V_0 as a space constituted by a set of translated elementary functions $\{\phi(t{-}n)\}$. Then,

$$s(t) \approx \sum_n c_{0,n}\phi(2^0 t - n) = \sum_n c_{0,n}\phi(t - n) \qquad (4.12)$$

where $c_{0,n}$ represents the weight of the function $\phi(t{-}n)$.

Similarly, the approximation plotted in (f) can be accomplished by translated piecewise constant function $\phi(2t)$ in Fig. 4–6e, that is

$$s(t) \approx \sqrt{2}\sum_n c_{1,n}\phi(2^1 t - n) = \sqrt{2}\sum_n c_{1,n}\phi(2t - n) \qquad (4.13)$$

where $\phi(2t)$ is dilated $\phi(t)$. It is interesting to note that $\phi(t)$ can be written in terms of $\phi(2t)$. For example,

$$\phi(t) = \phi(2t) + \phi(2t - 1) \qquad (4.14)$$

Because the interval of $\phi(2t)$ is half that of $\phi(t)$, the piecewise approximation illustrated in Fig. 4–6f has smaller error (or say better resolution) than that in Fig. 4–6b. We define V_1 as a space determined by a set of translated elementary functions $\{2^{1/2}\phi(2^1 t - n)\}$. Accordingly, V_m is a space determined by a set of translated elementary functions $\{2^{m/2}\phi(2^m t - n)\}$.

Obviously, as intervals of piecewise constant functions $2^{m/2}\phi(2^m t)$ get smaller and smaller, the resolution becomes better and better. When m goes to infinity, the approximation manifestly converges to $s(t)$. The factor $2^{m/2}$ ensures that the basis function $2^{m/2}\phi(2^m t{-}n)$ has unit energy. When m goes to minus infinity, the basis function $\phi_{m,n}(t)$ converges to zero.

It is important to note that *as long as the translations of* $\phi(t)$, $\{\phi(t{-}n)\}$, *are orthonormal, the translations of dilated* $\phi(t)$, *such as* $\{2^{m/2}\phi(2^m t{-}n)\}$ *for a fixed* m, *must also be orthonormal*, i.e.,

$$\int \phi_{m,n}(t)\phi^*_{m,n'}(t)dt = 2^m \int \phi(2^m t - n)\phi^*(2^m t - n')dt$$

$$= \int \phi(t-n)\phi^*(t-n')dt$$

$$= \delta(n-n') \qquad \forall\, m \in Z \tag{4.15}$$

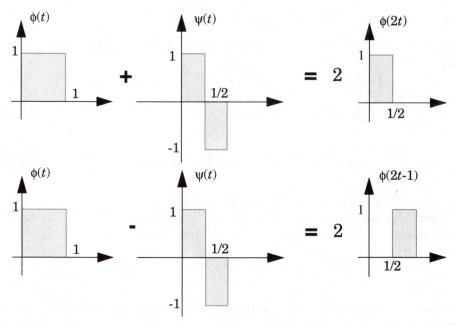

Fig. 4–7 The set of $\{\phi(2t-n)\}$ can be completely represented by the set of $\{\phi(t-n)\}$ plus $\{\psi(t-n)\}$.

The relation (4.14) shows that low resolution basis function $\phi(t)$ can be completely determined by high resolution basis $\phi(2t)$. In other words, $V_m \subset V_{m+1}$. Alternatively, as depicted in Fig. 4–7, we can also write high resolution basis $\phi(2t)$ by low resolution basis $\phi(t)$, plus some detailed information determined by $\{\psi(t-n)\}$. The function $\psi(t)$ is defined as

$$\psi(t) = \begin{cases} 1 & 0 \le t < \dfrac{1}{2} \\[2mm] -1 & \dfrac{1}{2} \le t < 1 \\[2mm] 0 & \text{otherwise} \end{cases} \tag{4.16}$$

Obviously, the translations of $\psi(t)$ are orthonormal, i.e.,

$$\int \psi(t-n)\psi^*(t-n')dt = \delta(n-n')$$

Moreover, $\psi(t)$ is orthogonal to $\phi(t)$, i.e.,

$$\int \phi(t-m)\psi^*(t-n)dt \equiv 0$$

From Fig. 4–7, we can see that

$$\phi(2t) = \frac{\phi(t)+\psi(t)}{2}$$

and

$$\phi(2t-1) = \frac{\phi(t)-\psi(t)}{2}$$

Let W_0 be a space that is constituted by the set of $\{\psi(t-n)\}$. Then, W_0 contains all translations of $\psi(t)$, $\psi(t-n)$.

One most important feature of piecewise approximation is that the high resolution approximation can be represented in terms of low level resolution plus some details. The reader can verify that

$$s(t) \approx \sqrt{2}\sum_n c_{1,n}\phi(2t-n) = \sum_n \frac{c_{1,2n}+c_{1,2n+1}}{2}\phi(t-n) + \frac{c_{1,2n}-c_{1,2n+1}}{2}\psi(t-n) \quad (4.17)$$

The second term is plotted in Fig. 4–6d, which is in W_0. The processing described by (4.17) is commonly remembered by

$$V_1 = V_0 \oplus W_0 \quad\quad\quad\quad\quad\quad (4.18)$$

which implies that the subspace W_0 is an orthogonal complementary space of V_0 in V_1. Obviously, $V_0 \subset V_1$ and $W_0 \subset V_1$. If continuing to carry over the decomposition in (4.18), then we have

$$V_1 = V_0 \oplus W_0 = V_{-1} \oplus W_{-1} \oplus W_0 = V_m \oplus W_m \ldots W_{-2} \oplus W_{-1} \oplus W_0 \quad (4.19)$$

where the space W_m is constituted by $\{\psi_{m,n}(t)\}$.

Because the translations of $\psi(t)$ are orthonormal, similar to (4.15), the translations of dilated $\psi(t)$, $\{2^{m/2}\psi(2^m t-n)\} = \{\psi_{m,n}(t)\}$ for a fixed m, must also be orthonormal. Moreover, since the space W_m is perpendicular to the space $W_{m'}$ for $m \neq m'$ as constructed before, $\psi_{m,n}(t)$ and $\psi_{m',n'}(t)$ are orthonormal to each other. That is,

$$\int \psi_{m,n}(t)\psi_{m',n'}(t)dt = 2^{(m+m')/2}\int \psi(2^m t-n)\psi^*(2^{m'}t-n')dt = \delta(m-m')\delta(n-n')$$

This relation obviously applies to the function $\psi(t)$ defined in (4.16). Fig. 4–8 depicts the case where $m = 0$ and $m' = 1$.

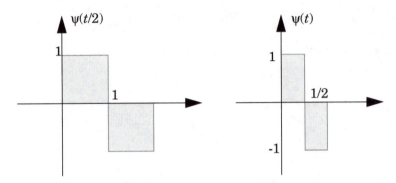

Fig. 4–8 $\{\psi_{m,n}(t)\}$ is orthonormal.

Note that $2^{m/2}\phi(2^{m}t)$ converges to zero as m goes to minus infinity. In the other words, V_m becomes the space with only zero when m approaches minus infinity. In this case, (4.28) reduces to

$$V_1 = \ldots \oplus W_{-2} \oplus W_{-1} \oplus W_0 \qquad (4.20)$$

The above decomposition can be generalized to

$$V_m = \ldots \oplus W_{-2} \oplus W_{-1} \oplus W_0 \oplus W_1 \oplus \ldots \oplus W_{m-1} \qquad \forall\, m \in Z \qquad (4.21)$$

As mentioned earlier, we can improve the accuracy of the approximation simply by reducing the interval of the function $2^{m/2}\phi(2^{m}t)$ (that is, increase m). Therefore,

$$2^{m/2}\sum_{n}c_{m,n}\phi(2^{m}t-n) \to s(t) \qquad \text{for } m \to \infty \qquad (4.22)$$

Applying (4.21) and (4.17), we can rewrite (4.22) as

$$s(t) = \sum_{m=-\infty}^{\infty}\sum_{n=-\infty}^{\infty}d_{m,n}2^{m/2}\psi(2^{m}t-n) \qquad (4.23)$$

which implies

$$V_\infty = \lim_{m \to \infty} V_m = \ldots \oplus W_{-2} \oplus W_{-1} \oplus W_0 \oplus W_1 \oplus \ldots \qquad (4.24)$$

which is equal to the whole signal space. Note that (4.23) has exactly the same form as the wavelet series in (4.10). $\psi(t)$ defined in (4.16) is actually a Haar wavelet, one of the most well-known wavelet functions. Because $\{\psi_{m,n}(t)\}$ is orthonormal, the wavelet series coefficients $d_{m,n}$ can be readily computed by the regular inner product process, such as

$$d_{m,n} = 2^{m/2} \int s(t) \psi^*(2^m t - n) dt \qquad (4.25)$$

In this section, we see that a Haar wavelet series can be achieved by the piecewise approximation. Although the Haar wavelet is not suitable for joint time-frequency analysis due to its poor frequency localizations, the concept behind piecewise constant approximation is very useful. Based on the piecewise constant approximation, in the next section, we shall introduce a more general method, multiresolution analysis, to design mother wavelets.

4.3 Multiresolution Analysis

In this section, we shall generalize the idea of the piecewise approximation discussed in the preceding section to the so-called *multiresolution analysis*, then derive a general algorithm to calculate the mother wavelet $\psi(t)$.

Let L^2 denote the whole signal space with all finite energy signals, that is,

$$\int |s(t)|^2 dt < \infty$$

Let $V_m, m \in Z$ be a nested subspace of L^2, i.e., $V_m \subset V_{m+1} \subset L^2$. The subspace sequence V_m is called a MRA (multiresolution analysis) of L^2 if the following holds:
(i) $s(t) \in V_m$ if and only if $s(2t) \in V_{m+1}$ for $m \in Z$;
(ii) The intersection of the sequence V_m has only zero signal, i.e.,

$$\bigcap_m V_m = \{0\}$$

(iii) Any signal in L^2 can be approximated by signals in the union of the spaces V_m;
(iv) There exists a signal $\phi(t) \in V_0$ such that its translation $\phi(t-n)$, $n \in Z$, forms an orthonormal basis of V_0, i.e.,

$$\int \phi(t) \phi^*(t-n) dt = \delta(n) \qquad (4.26)$$

and for any $s(t)$ in V_0,

$$s(t) = \sum_n c_n \phi(t-n)$$

where

$$c_n = \int s(t) \phi^*(t-n) dt$$

$\phi(t)$ is named a *scaling function*.

Obviously, the piecewise approximation example in the previous section is an MRA. The goal of MRA is to find the orthogonal complementary space W_0 of V_0 in V_1 and corresponding mother wavelet $\psi(t)$. Because the mother wavelet $\psi(t)$ is closely related to the scaling function $\phi(t)$, in what follows, we shall start with the scaling function $\phi(t)$ and then derive the mother wavelet $\psi(t)$.

By property (i), if $\phi(t/2) \in V_{-1}$ then $\phi(t/2) \in V_0$. From property (iv), there are constants h_n such that

$$\phi\left(\frac{t}{2}\right) = 2\sum_n h_n \phi(t - n) \tag{4.27}$$

which is named the *dilation equation*. In the case of the piecewise approximation example, $h_0 = h_1 = 1/2$, as described in (4.14).

Taking the Fourier transform in both sides of (4.27) yields

$$\int \phi\left(\frac{t}{2}\right) e^{-j\omega t} dt = 2\sum_n h_n \int \phi(t - n) e^{-j\omega t} dt \tag{4.28}$$

By replacing the integration variables, we can rewrite (4.28) as

$$2\int \phi(t) e^{-j2\omega t} dt = 2\sum_n h_n \int \phi(t) e^{-j\omega(t + n)} dt = 2\sum_n h_n e^{-j\omega n} \int \phi(t) e^{-j\omega t} dt \tag{4.29}$$

In other words,

$$\Phi(2\omega) = H(\omega)\Phi(\omega) \tag{4.30}$$

or

$$\Phi(\omega) = H\left(\frac{\omega}{2}\right)\Phi\left(\frac{\omega}{2}\right) \tag{4.31}$$

where $\Phi(\omega)$ denotes the Fourier transform of $\phi(t)$. $H(\omega)$ is the discrete Fourier transform of h_n, which is periodic in frequency. As long as $\Phi(0) \neq 0$, $H(0) = 1$. This means that $H(\omega)$ is a lowpass filter. If we continue to carry out such decomposition, then

$$\Phi(\omega) = H\left(\frac{\omega}{2}\right)\Phi\left(\frac{\omega}{2}\right) = H\left(\frac{\omega}{2}\right)H\left(\frac{\omega}{4}\right)\Phi\left(\frac{\omega}{4}\right) = \prod_{k=1}^{\infty} H\left(\frac{\omega}{2^k}\right)\Phi(0) \tag{4.32}$$

Without loss of generality, let $\Phi(0) = 1$, that is,

$$\Phi(0) = \int \phi(t) dt = 1 \tag{4.33}$$

the normalized scaling function $\phi(t)$. Substituting (4.33) into (4.32) yields

$$\Phi(\omega) = \prod_{k=1}^{\infty} H\left(\frac{\omega}{2^k}\right) \tag{4.34}$$

which shows that as long as it satisfies the dilation equation (4.27), the scaling function $\phi(t)$ is completely determined by the lowpass type filter $H(\omega)$. According to the MRA definition, the translations of $\phi(t)$ are orthonormal (condition (iv)). The next question is what this condition implies to $H(\omega)$.

Because $\{\phi(t-k)\}$ is orthogonal, that is,

$$\int \phi(t)\phi^*(t-n)dt = \delta(n) \qquad n \in Z$$

by the Parseval's relationship, we have

$$\int \Phi(\omega)\Phi^*(\omega)e^{-jn\omega}d\omega = 2\pi\delta(n) \tag{4.35}$$

Taking the summation at both sides with respect to n, we obtain

$$\sum_n \int \Phi(\omega)\Phi^*(\omega)e^{-j\omega n}d\omega = 2\pi \tag{4.36}$$

That is,

$$\int \Phi(\omega)\Phi^*(\omega)\sum_n e^{-j\omega n}d\omega = 2\pi\int\Phi(\omega)\Phi^*(\omega)\sum_n \delta(\omega-2n\pi)d\omega = 2\pi \tag{4.37}$$

Therefore,

$$\sum_k |\Phi(\omega+2k\pi)|^2 = 1 \qquad \forall\, \omega \tag{4.38}$$

Substituting (4.31) into (4.38) yields

$$\sum_k |\Phi(\omega+2k\pi)|^2 = \sum_k \left|H\left(\frac{\omega+2k\pi}{2}\right)\Phi\left(\frac{\omega+2k\pi}{2}\right)\right|^2$$

$$= \sum_k \left|H\left(\frac{\omega}{2}+k\pi\right)\Phi\left(\frac{\omega}{2}+k\pi\right)\right|^2$$

$$= \sum_{n=-\infty}^{\infty} \left|H\left(\frac{\omega}{2}+2n\pi\right)\Phi\left(\frac{\omega}{2}+2n\pi\right)\right|^2 + \sum_{n=-\infty}^{\infty} \left|H\left(\frac{\omega}{2}+(2n+1)\pi\right)\Phi\left(\frac{\omega}{2}+(2n+1)\pi\right)\right|^2$$

$$= 1 \tag{4.39}$$

where we partition the variable k into even and odd parts. Because $H(\omega)$ is periodic in frequency, that is, $H(\omega) = H(\omega+2\pi)$, the identity (4.39) reduces to

$$\left|H\left(\frac{\omega}{2}\right)\right|^2 \sum_{n=-\infty}^{\infty} \left|\Phi\left(\frac{\omega}{2} + 2n\pi\right)\right|^2 + \left|H\left(\frac{\omega}{2} + \pi\right)\right|^2 \sum_{n=-\infty}^{\infty} \left|\Phi\left(\frac{\omega}{2} + (2n+1)\pi\right)\right|^2 = 1 \qquad (4.40)$$

By the relation (4.38), the identity (4.40) becomes

$$H(\omega)H^*(\omega) + H(\omega+\pi)H^*(\omega+\pi) = 1 \qquad \forall \omega \qquad (4.41)$$

As shown earlier, $H(0) = 1$, therefore (4.41) implies that $H(\pi) = 0$.

So far, we have proved that the orthonormal scaling function $\phi(t)$ can be generated by a lowpass filter $H(\omega)$, in which $H(0) = 1$ and $H(\pi) = 0$. To compute the mother wavelet, let's introduce another function $G(\omega)$ so that

$$H(\omega)G^*(\omega) + H(\omega+\pi)G^*(\omega+\pi) = 0 \qquad (4.42)$$

The pair of $H(\omega)$ and $G(\omega)$ is named *quadrature mirror filters* for MRA. One solution of (4.42) is

$$G(\omega) = -e^{-j\omega}H^*(\omega+\pi) \qquad (4.43)$$

Substituting $H(0) = 1$ and $H(\pi) = 0$ into (4.43) yields $G(0) = 0$ and $G(\pi) = 1$, respectively. This means that $G(\omega)$ is a highpass filter. From (4.43), the inverse Fourier transform of $G(\omega)$ can be computed as

$$g_k = (-1)^k h_{1-k} \qquad (4.44)$$

Assume that $\psi(t)$ is a function whose Fourier transform $\Psi(\omega)$ satisfies

$$\Psi(\omega) = G\left(\frac{\omega}{2}\right)\Phi\left(\frac{\omega}{2}\right) = G\left(\frac{\omega}{2}\right)\prod_{k=2}^{\infty} H\left(\frac{\omega}{2^k}\right) \qquad (4.45)$$

Then the corresponding time relationship is

$$\psi(t) = 2\sum_k g_k \phi(2t-k) \qquad (4.46)$$

One can prove (see [36] and [121]) that under minor conditions on $H(\omega)$, $\psi(t-n)$ for all integers n form an orthonormal basis for the orthogonal complementary space W_0 of V_0 in V_1, i.e., $V_1 = V_0 \oplus W_0$.

Because the translations of $\psi(t)$, $\{\psi(t-n)\}$, forms an orthonormal space W_0, similar to (4.15), the translations of dilated $\psi(t)$, $\{\psi_{m,n}(t)\}$ for a fixed m, must also form an orthonormal basis for the orthogonal complimentary space W_m of V_m in V_{m+1}. By properties (i) to (iii),

$$L^2 = \ldots \oplus W_{m-1} \oplus W_m \oplus \ldots$$

Therefore, the dilations and translations $\psi_{m,n}(t)$ of $\psi(t)$ form an orthonormal basis for the signal space L^2. This means that for any signal $s(t)$, we have

$$s(t) = \sum_{m=-\infty}^{\infty} \sum_{n=-\infty}^{\infty} d_{m,n} \, \psi_{m,n}(t)$$

where

$$d_{m,n} = \int s(t) \psi^*_{m,n}(t) dt$$

which is exactly what we want. Therefore, $\psi(t)$ is the mother wavelet. The constants $d_{m,n}$ are *wavelet series coefficients*.

Example 4-1   Haar wavelet

$$H(\omega) = \frac{1}{2}(1 + \exp\{-j\omega\}) \tag{4.47}$$

Because $H(0) = 1$ and $H(\pi) = 0$, $H(\omega)$ in (4.47) is lowpass. From (4.32), the Fourier transform of the scaling function is

$$\Phi(\omega) = \prod_{k=1}^{\infty} H\left(\frac{\omega}{2^k}\right) = \prod_{k=1}^{\infty} \frac{1}{2}\left(1 + \exp\left\{-j\frac{\omega}{2^k}\right\}\right)$$

$$= \prod_{k=1}^{\infty} \exp\left\{-j\frac{\omega}{2^{k+1}}\right\}\frac{1}{2}\left(\exp\left\{j\frac{\omega}{2^{k+1}}\right\} + \exp\left\{-j\frac{\omega}{2^{k+1}}\right\}\right)$$

$$= \prod_{k=1}^{\infty} \exp\left\{-j\frac{\omega}{2^{k+1}}\right\}\cos\left(\frac{\omega}{2^{k+1}}\right)$$

$$= \exp\left\{-j\omega \sum_{k=1}^{\infty} \frac{1}{2^{k+1}}\right\}\prod_{k=1}^{\infty} \cos\left(\frac{\omega}{2^{k+1}}\right)$$

$$= \exp\{-j\omega/2\}\prod_{k=1}^{\infty} \cos\left(\frac{\omega}{2^{k+1}}\right)$$

Because[*]

$$\prod_{k=1}^{\infty} \cos\left(\frac{\omega}{2^{k+1}}\right) = \frac{\sin\omega/2}{\omega/2}$$

___

[*] See formula 1.439, p. 38, "Table of Integrals, Series, and Products," by I.S. Gradshteyn and I.M. Ryshik, Academic Press, New York and London, 1965.

we have

$$\Phi(\omega) = \exp\{-j\omega/2\}\frac{\sin\omega/2}{\omega/2} \tag{4.48}$$

From (4.43), the highpass filter corresponding to $H(\omega)$ in (4.47) is

$$G(\omega) = -\exp\{-j\omega\}H^*(\omega+\pi) = \frac{1}{2}(1 - \exp\{-j\omega\}) \tag{4.49}$$

Obviously, $H(\omega)$ and $G(\omega)$ constitute quadrature mirror filters, defined earlier. By (4.45) and (4.49), we can compute the Fourier transform of the wavelet, i.e.,

$$\Psi(\omega) = G\left(\frac{\omega}{2}\right)\Phi\left(\frac{\omega}{2}\right) = \frac{1}{2}\left(1 - \exp\left\{-j\frac{\omega}{2}\right\}\right)\exp\left\{-j\frac{\omega}{4}\right\}\frac{\sin\omega/4}{\omega/4}$$

$$= \frac{1}{2}\left(\exp\left\{-j\frac{\omega}{4}\right\} - \exp\left\{-j\frac{3}{4}\omega\right\}\right)\frac{\sin\omega/4}{\omega/4} \tag{4.50}$$

The magnitude is

$$|\Psi(\omega)| = \sqrt{2\left(1 - \cos\frac{\omega}{2}\right)\left(\frac{2\sin\omega/4}{\omega}\right)^2} = \sqrt{(2\sin\omega/4)^2\left(\frac{2\sin\omega/4}{\omega}\right)^2}$$

$$= \frac{(\sin\omega/4)^2}{|\omega/4|} \tag{4.51}$$

which is sketched in Fig. 4–9.

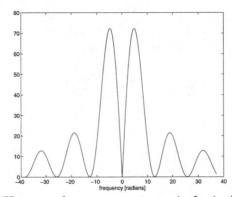

Fig. 4–9 $|\Psi(\omega)|$. (Haar wavelet possesses strong ripples in the frequency domain.)

The inverse Fourier transform of $\Psi(\omega)$ is

$$\psi(t) = \begin{cases} 1 & 0 \leq t < \dfrac{1}{2} \\ -1 & \dfrac{1}{2} \leq t < 1 \\ 0 & \text{otherwise} \end{cases} \qquad (4.52)$$

which is exactly the Haar wavelet. The corresponding time domain scaling function and Haar wavelet are illustrated in Fig. 4–10. Although the Haar wavelet is compactly supported in time, it has strong ripples in the frequency domain. Therefore, the Haar wavelet is not suitable for joint time-frequency analysis.

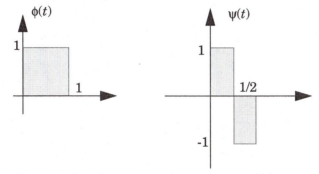

Fig. 4–10 Haar scaling function $\phi(t)$ and wavelet $\psi(t)$. (Haar wavelet is compactly supported in the time domain.)

By a similar procedure, the reader can compute a mother wavelet based on the ideal lowpass filter

$$H(\omega) = \begin{cases} 1 & \omega \leq \left|\dfrac{\pi}{2}\right| \\ 0 & \text{otherwise} \end{cases} \qquad (4.53)$$

The resulting wavelet is called a *sinc wavelet*.

Except for a few cases, however, the analytical solutions of $\phi(t)$ and $\psi(t)$ for a given lowpass filter in general do not exist. Very often, $\phi(t)$ and $\psi(t)$ have to be determined numerically. Fig. 4–11 illustrates a Daubechies wavelet and corresponding scaling function, in which the lowpass filter is

$$h_k = \frac{1+\sqrt{3}}{8}, \frac{3+\sqrt{3}}{8}, \frac{3-\sqrt{3}}{8}, \frac{1-\sqrt{3}}{8}$$

The main features of Daubechies wavelets are that they are smooth and compactly supported in time domain. As the order of the lowpass filter gets larger, the wavelets tends to be smoother. At the same time, the time support region becomes wider.

In short, both scaling function $\phi(t)$ and mother wavelet $\psi(t)$ can be generated from a pair of quadrature mirror filters $H(\omega)$ and $G(\omega)$. In addition to $H(\pi) = G(0) = 0$ and $H(0) = G(\pi) = 1$, we can impose other conditions to lowpass filters. The most basic requirement for the scaling and mother wavelet is smoothness. This is usually achieved by imposing the vanishing of derivatives of lowpass filter $H(\omega)$ at π up to a certain order. The reader can find excellent descriptions of quadrature mirror filter design in [168], [179], and [180].

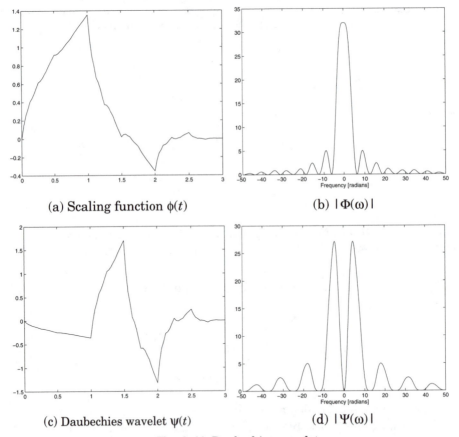

(a) Scaling function $\phi(t)$

(b) $|\Phi(\omega)|$

(c) Daubechies wavelet $\psi(t)$

(d) $|\Psi(\omega)|$

Fig. 4–11 Daubechies wavelet

4.4 Discrete Wavelet Transformation and Digital Filter Banks

In the last three sections, we introduce the wavelet transformation for continuous-time signals. As a matter of fact, the majority of signals encountered in our everyday life are the functions of discrete-time. Therefore, it is important for us to investigate the wavelet transform for discrete-time samples. The main issue here is that the discrete version of a wavelet cannot be obtained by simply replacing t with the discrete-time index i (4.10). This is because the analytic mother wavelets $\psi(t)$ do not exist in general. Moreover, the quantity $2^m i$ does not remain an integer for $m < 0$. Hence, special care has to be taken when we develop the discrete wavelet transformation.

Based upon our previous discussions, we could conclude that if the signal $s(t)$ is in V_k for finite k, then $s(t)$ should be completely determined by

$$s(t) = \sum_{n=-\infty}^{\infty} c_{k,n} \phi_{k,n}(t) \tag{4.54}$$

Because $V_k = V_{k-1} \oplus W_{k-1}$, (4.54) can be rewritten as

$$s(t) = \sum_n c_{m_0,n} \phi_{m_0,n}(t) + \sum_{m=m_0}^{k-1} \sum_n d_{m,n} \psi_{m,n}(t) \qquad k > m_0 \tag{4.55}$$

Coefficients $c_{m,n}$ and $d_{m,n}$ are inner products between $s(t)$ and $\phi_{m,n}(t)$, and $\psi_{m,n}(t)$, respectively. Therefore, the constants $d_{m,n}$ are *wavelet series coefficients*.

By the Parseval's equality, we have

$$c_{k,n} = 2^{k/2} \int s(t) \phi(2^k t - n) dt = \frac{1}{2\pi} 2^{-k/2} \int S(\omega) \Phi^*(2^{-k}\omega) e^{-j2^{-k}\omega n} d\omega \tag{4.56}$$

For a normalized $\phi(t)$ (see (4.33)), that is, $\Phi(0) = 1$, (4.56) reduces to

$$c_{k,n} \approx \frac{1}{2\pi} 2^{-k/2} \int S(\omega) e^{-j2^{-k}n\omega} d\omega = 2^{-k/2} s\left(\frac{n}{2^k}\right) \tag{4.57}$$

for a large k. This means that $c_{k,n}$ is approximately equal to the sample of $s(t)$ at $t = 2^{-k}n$ with a scale $2^{-k/2}$. The higher the resolution is (that is, larger k), the smaller the error. The reader can find detailed error analysis in [192] and [193].

Without loss of generality, let

$$c_{k,n} \equiv s[n] \equiv s(t)\big|_{t=2^{-k}n} \tag{4.58}$$

Because

$$c_{k-1,n} = \int s(t) \phi^*_{k-1,n}(t) dt = 2^{(k-1)/2} \int s(t) \phi^*\left(\frac{2^k t - 2n}{2}\right) dt$$

$$= 2^{(k-1)/2} \int s(t) 2 \sum_i h_i \phi^*(2^k t - 2n - i) dt \tag{4.59}$$

where we use the relation described in (4.27). By exchanging the summation and integration, (4.59) becomes

$$c_{k-1,n} = \sqrt{2}\sum_i h_i \int s(t)\phi^*_{k,2n+i}(t)dt = \sqrt{2}\sum_i h_i c_{k,2n+i} = \sqrt{2}\sum_i h_{i-2n}c_{k,i} \quad (4.60)$$

which implies that once $c_{k,n}$ is known, we can recursively compute $c_{m,n}$, for $m <$ k, by a lowpass filter $H(\omega)$. Fig. 4–12 illustrates the operation in (4.60), where the block following the lowpass filter denotes downsampling by two.

$$c_{k,n} \longrightarrow \boxed{H^*(\omega)} \longrightarrow \boxed{\downarrow 2} \longrightarrow c_{k-1,n}$$

Fig. 4–12 Low resolution coefficients $c_{k-1,n}$ can be recursively computed by lowpassing high resolution coefficients $c_{k,n}$.

Similarly, we can prove that

$$d_{k-1,n} = \sqrt{2}\sum_i g_{i-2n}c_{k,i} \quad (4.61)$$

where g_k is defined in (4.44), which is a highpass filter. Note that $d_{m,n}$ are wavelet series coefficients. Eqs.(4.60) and (4.61) imply that $d_{m,n}$ can be obtained by filter banks, as illustrated in Fig. 4–13. The output of highpass filters are wavelet series coefficients $d_{m,n}$. It is interesting to note that for discrete-time samples, the wavelet transform can be accomplished by directly applying filter banks, without computing mother wavelet function $\psi(t)$.

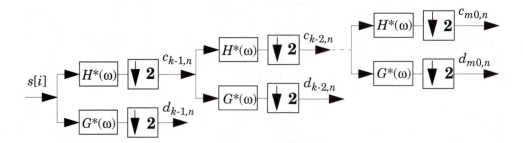

Fig. 4–13 Implementation of discrete wavelet transform via digital filter banks.

The relations (4.60) and (4.61) show that, given high-resolution coefficients, we can directly compute the low-resolution coefficients. Conversely, we can compute the high-resolution coefficients based upon the low-resolution coefficients, i.e.,

$$c_{m,n} = \sqrt{2}\left(\sum_i h_{n-2i} c_{m-1,i} + \sum_i g_{n-2i} d_{m-1,i}\right) \tag{4.62}$$

which implies that we can recover the original signal $s[i]$ by filter banks, as illustrated in Fig. 4–14.

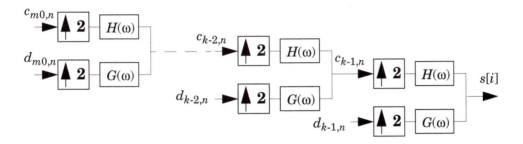

Fig. 4–14 The original sample can be recovered by digital filter banks.

PROOF

From (4.60) and (4.61), the right side of (4.62) can be written as

$$\sqrt{2}\left(\sum_i h_{n-2i} c_{m-1,i} + \sum_i g_{n-2i} d_{m-1,i}\right)$$

$$= 2\left(\sum_i h_{n-2i}\sum_k h_{k-2i} c_{m,k} + \sum_i g_{n-2i}\sum_k g_{k-2i} c_{m,k}\right)$$

$$= 2\sum_k c_{m,k}\sum_i (h_{n-2i} h_{k-2i} + g_{n-2i} g_{k-2i})$$

$$= 2\sum_k c_{m,k}\sum_i (h_{n-2i} h_{k-2i} + (-1)^{(n+k)} h_{1-n+2i} h_{1-k+2i}) \tag{4.63}$$

where we use the relation described in (4.44). In order to have (4.62), we have to prove that

$$2\sum_i (h_{n-2i} h_{k-2i} + (-1)^{(n+k)} h_{1-n+2i} h_{1-k+2i}) = \delta(k-n) \tag{4.64}$$

There are two cases: $n+k$ is odd and $n+k$ is even. When $n+k$ is odd, i.e., $n+k = 2p+1$, for an integer p, the left side of (4.64) reduces to

$$2\sum_i (h_{n-2i}h_{(2p+1)-n-2i} - h_{1-n+2i}h_{1-(2p+1)+n+2i})$$

$$= 2\sum_i h_{n-2i}h_{2p+1-n-2i} - 2\sum_i h_{n+2i-2p}h_{2i-n+1}$$

$$= 2\sum_i h_{2i+n}h_{2i+2p+1-n} - 2\sum_i h_{2i+n}h_{2i+2p+1-n} = 0 \tag{4.65}$$

When $n+k$ is even, i.e., $n+k = 2p$, for an integer p, the left side of (4.65) reduces to

$$2\sum_i (h_{n-2i}h_{2p-n-2i} + h_{1-n+2i}h_{1-2p+n+2i})$$

$$= 2\sum_i h_{2i+n}h_{2i+2p-n} + 2\sum_i h_{2i+1-n}h_{2(i-p)+1+n}$$

$$= 2\sum_i h_{2i+n}h_{2i+2p-n} + 2\sum_i h_{2i+2p+1-n}h_{2i+1+n}$$

$$= 2\sum_i h_{2i+n}h_{2i+2p-n} + 2\sum_i h_{2i+1+2p-n}h_{2i+1+n} \tag{4.66}$$

Because $2i$ is even and $2i+1$ is odd, we can group two summations in (4.66) into

$$2\sum_i h_{i+n}h_{i+2p-n} = 2\sum_i h_i h_{i+2p-2n} = 2\sum_i h_i h_{i+2(p-n)} \tag{4.67}$$

Because the lowpass filter $H(\omega)$ satisfies Eq.(4.41), i.e.,

$$H(\omega)H^*(\omega) + H(\omega+\pi)H^*(\omega+\pi) = 1$$

Taking the inverse Fourier transform at both sides yields

$$2\sum_i h_i h_{i-2n} = \delta(n) \tag{4.68}$$

for all integers n. Substituting (4.68) into (4.67) leads to

$$2\sum_i h_i h_{i+2(p-n)} = \delta(p-n) = \delta(k-n) \tag{4.69}$$

Hence,

$$c_{m,n} = \sqrt{2}\left(\sum_i h_{n-2i}c_{m-1,i} + \sum_i g_{n-2i}d_{m-1,i}\right)$$

Summary

In this section, we introduced the concept of wavelet transformation, which is closely related to the conventional constant Q analysis. Unlike the STFT, where we can virtually use any function as a window function, to a complete reconstruction the mother wavelet has to satisfy certain conditions. Therefore, one fundamental issue in wavelet analysis is the selection of the mother wavelet $\psi(t)$. For the sake of simplicity, we focused our discussions on orthonormal wavelets[*]. In these cases, the mother wavelet can be generated via multiresolution analysis structure. That is, from Eq.(4.45)

$$\Psi(\omega) = G\left(\frac{\omega}{2}\right)\Phi\left(\frac{\omega}{2}\right) = G\left(\frac{\omega}{2}\right)\prod_{k=2}^{\infty} H\left(\frac{\omega}{2^k}\right)$$

where $\Phi(\omega)$ denotes the scaling function. $H(\omega)$ and $G(\omega)$ represent a pair of lowpass and highpass filters of quadrature mirror filters. Consequently, the problem of selecting the mother wavelet reduces to the design of quadrature mirror filters. For a discrete-time signal, we don't even need to explicitly calculate the mother wavelet $\psi(t)$. The wavelet transform $d_{m,n}$ could be well approximated by digital filter banks.

As we know, wavelets is a very broad topic. What is presented in this chapter is the most fundamental. We only address the orthonormal bases implemented by means of digital filter banks. The reader can find more comprehensive materials in [24], [35], [168], [180], and [186].

---

[*] The theory developed in this chapter could be generalized to biorthogonal wavelets. For example, the biorthogonal wavelets and corresponding wavelet coefficients could also be approximated by the digital filter banks.

Wigner-Ville Distribution

*T*he representations that describe a signal's frequency behavior fall predominantly into two categories: linear representations such as the Fourier transform and quadratic representations such as the power spectrum. Previously we described the linear joint time-frequency representations, the short-time Fourier transform, the Gabor expansion (which can be considered an inverse of a sampled short-time Fourier transform), as well as wavelets. In Chapters 5, 6, 7, and 8, we will introduce the counterpart to the power spectrum: the quadratic, or bilinear, joint time-frequency representation. Although dozens of bilinear joint time-frequency representations have been proposed over the last five decades, we shall start with the Wigner-Ville distribution because it is simple and powerful.

In Sections 5.1 and 5.2, we discuss the motivation and general properties of the Wigner-Ville distribution. What makes the Wigner-Ville distribution so unique are its descriptions of a signal's time-varying nature better than many other representations, such as STFT spectrogram and scalogram (the square of the wavelet transform), which are known in the area of joint time-frequency analysis. Moreover, the Wigner-Ville distribution possesses many properties useful for signal analysis. The problems of the Wigner-Ville distribution have been the so-called cross-term[*] interference that severely limits the applications of the

---

[*] The cross-term in fact is not an appropriate phrase to describe the unwanted terms appearing in the Wigner-Ville distribution, though it is used in almost all literature. We shall clarify it in Chapter 7.

Wigner-Ville distribution. Therefore, a major research effort has existed to find time-frequency representations that preserve the properties of the Wigner-Ville distribution but have reduced cross-term interference. Such a joint time-frequency representation is the focus of the rest of Part 2. In Section 5.3, we investigate the cause and effect of the cross-term interference. In Section 5.4, we address the relationships between Wigner-Ville distribution and STFT spectrogram and scalogram, respectively. Section 5.5 is devoted to the smoothed Wigner-Ville distribution and Wigner-Ville distribution of analytic signals. Both of them are traditionally used for cross-term interference reduction. Although they have certain limitations, they are simple and effective for many real applications. Finally, in Section 5.6, we discuss the numerical implementation issue of the Wigner-Ville distribution.

Wigner-Ville distribution has been known for many years. What we discuss in this chapter are only those aspects which are closely related to our future developments. The reader can find more comprehensive treatments of Wigner-Ville distribution in [25], [133], [134], and [135].

5.1   Time-Dependent Power Spectrum

The square of the Fourier transform is called the *power spectrum*, which characterizes the signal's energy distribution in the frequency domain. While the Fourier transform is linear, the power spectrum is the quadratic function of frequencies. Accordingly, we also use the square of short-time Fourier transform and wavelet transform to describe the signal's energy distribution in joint time-frequency domain. The squares of short-time Fourier transform and wavelet transform are named STFT spectrogram and scalogram, respectively. As discussed in previous chapters, the results obtained from STFT spectrogram and scalogram are subject to the selection of analysis functions. To overcome those problems, in the present chapter we shall introduce a more general method of describing the signal's energy distribution in joint time-frequency domain.

According to the Wiener-Khinchin theorem, the power spectrum can also be considered as the Fourier transform of the auto-correlation function $R(\tau)$, i.e.,

$$\text{PS}(t, \omega) = |S(\omega)|^2 = \int R(\tau) \exp\{-j\omega\tau\} d\tau \qquad (5.1)$$

where $R(\tau)$ is computed by

$$R(\tau) = \int s(t) s^*(t - \tau) dt \qquad (5.2)$$

Eq. (5.1) is not a function of time, which indicates how much energy is present in frequency ω over the entire time period. But it does not show how the spectrum is distributed in time. Based on (5.1), there is no way to tell whether or not a signal's power spectrum changes over time. Therefore, the standard power spectrum is inadequate to depict signals whose frequency contents evolve with time, such as most biomedical signals, speech signals, stock indexes, and vibration signals.

By examining (5.1), we can see that one possible way to depict a time-dependent spectrum is to make the auto-correlation function time-dependent. The resulting Fourier transform of the time-dependent auto-correlation function $R(t,\tau)$, with respect to variable τ, is then a function of time, i.e.,

$$P(t, \omega) = \int R(t, \tau) \exp\{-j\omega\tau\} d\tau \tag{5.3}$$

We name $P(t,\omega)$ a *time-dependent power spectrum*. The question that remains is how to determine the time-dependent auto-correlation function $R(t,\tau)$.

Apparently, the choice of $R(t,\tau)$ is not arbitrary. For example, because $P(t,\omega)$ presumably describes the time-dependent power spectrum, adding all instantaneous-time spectrum $P(t_0,\omega)$ should yield the total power spectrum $|S(\omega)|^2$, i.e.,

$$\int P(t, \omega) dt = |S(\omega)|^2 \tag{5.4}$$

which is traditionally called the *frequency marginal condition*. Conversely, the integration along the frequency axis should be equal to the instantaneous energy, i.e.,

$$\frac{1}{2\pi} \int P(t, \omega) d\omega = |s(t)|^2 \tag{5.5}$$

which is commonly known as the *time marginal condition*. If $P(t,\omega)$ represents signal energy distribution in the joint time-frequency domain, then we hope it is real-valued, that is,

$$P(t, \omega) = P^*(t, \omega) \tag{5.6}$$

From the conventional energy concept, we also wish that the time-dependent spectrum would be non-negative.

Most importantly, however, we need to ensure that $P(t,\omega)$ indeed identifies the signal's frequency content changes. This is the primary motivation of joint time-frequency analysis and it also is the most difficult property to justify. For linear representations, such as short-time Fourier transform and wavelet transform, the goodness of the representation can simply be judged by the concentration of elementary functions (or the finesse of tick marks). The higher the elementary functions are concentrated, the better the proposed representation describes a signal's local behaviors. Unfortunately, it is not the case for the time-dependent spectrum. For most time-dependent spectra, there are no explicit elementary functions. We shall introduce a criterion that is widely employed for signal processing after examining some examples in the next section.

The most popular time-dependent spectrum is STFT spectrogram. In addition to being the square of short-time Fourier transform, STFT spectrogram can also be written as the Fourier transform of time-dependent auto-correlation function $R(t,\tau)$, where

$$R(t, \tau) = \frac{1}{2\pi}\int A_s(\vartheta, \tau)A_\gamma(\vartheta, \tau)\exp\{j\vartheta t\}d\vartheta \tag{5.7}$$

where $A_s(\vartheta,\tau)$ represents the ambiguity functions of signal $s(t)$. $A_\gamma(\vartheta,\tau)$ is the ambiguity function of the analysis window $\gamma(t)$. Let's leave the proof of (5.7) until Chapter 6. What the reader should bear in mind is that STFT spectrogram only satisfies (5.6). It does not satisfy frequency marginal (5.4) and time marginal (5.5) conditions. The resolution of the STFT spectrogram is subject to the selection of the analysis window function $\gamma(t)$.

Much research over the last five decades has focused on the time-dependent spectrum. The methodology of discovering the desired time-dependent auto-correlation functions $R(t,\tau)$ has been a major development in this area. The reader can find a comprehensive review of this subject in [32]. We shall not delve into those details here, but rather introduce the results that have been well established, such as the Wigner-Ville distribution.

The Wigner[*] distribution [187] was originally developed for the area of quantum mechanics in 1932 and was introduced for signal analysis by a French scientist Ville [181] 15 years later. It is now commonly known in the signal processing community as the *Wigner-Ville distribution* (WVD).

In the WVD, the time-dependent auto-correlation function is chosen to be

$$R(t, \tau) = s\left(t + \frac{\tau}{2}\right)s^*\left(t - \frac{\tau}{2}\right) \tag{5.8}$$

Substituting the above time-dependent auto-correlation into (5.3) yields

$$\mathrm{WVD}_s(t, \omega) = \int s\left(t + \frac{\tau}{2}\right)s^*\left(t - \frac{\tau}{2}\right)\exp\{-j\omega\tau\}d\tau \tag{5.9}$$

Eq.(5.9) is usually called the auto-WVD. Accordingly, the cross-WVD is defined as

$$\mathrm{WVD}_{s,g}(t, \omega) = \int s\left(t + \frac{\tau}{2}\right)g^*\left(t - \frac{\tau}{2}\right)\exp\{-j\omega\tau\}d\tau \tag{5.10}$$

---

[*] Eugene P. Wigner was born on November 17, 1902, in Budapest, Hungary. He attended Budapest Lutheran High School, where he met John von Neumann (1903-1957). At age 23, Wigner received his doctorate in chemical engineering. He first came to the United States in 1930 as a lecturer in mathematics at Princeton, where he spent most of his career. He worked on a series of projects during World War II, including the Manhattan Project. On December 10, 1963, Eugene Wigner, with Maria Goeppert Mayer and J. Hans D. Jensen, received the Nobel Prize for their discoveries concerning the theory of the atomic nucleus and elementary particles, which were based on atomic research that had been conducted during the first three decades of the twentieth century. Wigner laid the groundwork for the revision of concepts concerning right-left symmetry by Chen Ning Yang and Tsung-Dao Lee, who won the Nobel Prize in 1957. Wigner's sister, Margaret married Paul Dirac, the brilliant Nobel laureate, whom Wigner always described as "my famous brother-in-law."

where $s(t)$ and $g(t)$ denote two different signals. It is easier to verify that

$$\text{WVD}_{s,g}(t, \omega) = \text{WVD*}_{g,s}(t, \omega) \qquad (5.11)$$

Consequently,

$$\text{WVD}_s(t, \omega) = \text{WVD*}_s(t, \omega) \qquad (5.12)$$

which implies that the auto-WVD is real-valued.

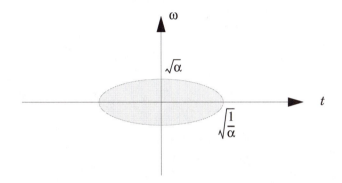

Fig. 5–1 The WVD of the Gaussian function is concentrated at (0,0). The parameter α controls the spread of WVD in time and frequency domains. The area of the ellipse where the levels are down to e^{-1} of their peak value is $A = \pi$, which is half the size of STFT.

Example 5–1   WVD with Gaussian-type analysis
                function

$$s(t) = \left(\frac{\alpha}{\pi}\right)^{\frac{1}{4}} \exp\left\{-\frac{\alpha}{2}t^2\right\} \qquad (5.13)$$

$s(t)$ is a normalized Gaussian function which has a unit energy. The corresponding WVD is

$$\text{WVD}_s(t, \omega) = \sqrt{\frac{\alpha}{\pi}} \int \exp\left\{-\frac{\alpha}{2}\left[\left(t + \frac{\tau}{2}\right)^2 + \left(t - \frac{\tau}{2}\right)^2\right]\right\} \exp\{-j\omega\tau\} d\tau$$

$$= \exp\{-\alpha t^2\} \sqrt{\frac{\alpha}{\pi}} \int \exp\left\{-\frac{\alpha}{4}\tau^2\right\} \exp\{-j\omega\tau\} d\tau$$

$$= 2\exp\left\{-\left[\alpha t^2 + \frac{1}{\alpha}\omega^2\right]\right\} \qquad (5.14)$$

which indicates that the WVD of the Gaussian function is concentrated at the origin (0,0), a signal's time and frequency center. The parameter α controls the spread of

WVD in time and frequency domain. A larger value of α leads to less spread in the time domain but large spread in the frequency domains, or vice versa. The contour plot of the WVD consists of concentric ellipses. The contour for the case where the levels are down to e^{-1} of their peak value is the ellipse indicated in Fig. 5–1. The area of this particular level ellipse is π. Compared to the STFT spectrogram in Example 3–1, the resolution of WVD is fixed, in which there is no window effect. For the STFT, the minimum area of the ellipse where the levels are down to e^{-1} of their peak value is $A = 2\pi$, which is twice as big as that of the WVD. In other words, the resolution of WVD is twice as good as that of a STFT spectrogram, if we use the area A as a measure. Moreover, the reader can easily verify that (5.14) satisfies both time and frequency marginal conditions.

In addition to the formulae (5.9) and (5.10), the WVD can also be computed from the frequency domain. Let

$$s_1(\tau) = s\left(t + \frac{\tau}{2}\right) \qquad \text{and} \qquad g_1(\tau) = g^*\left(t - \frac{\tau}{2}\right)$$

Then, the corresponding Fourier transforms are

$$s_1(\tau) \leftrightarrow S_1(\omega) = 2S(2\omega)e^{j2\omega t} \qquad \text{and} \qquad g_1(\tau) \leftrightarrow G_1(\omega) = 2G^*(2\omega)e^{-j2\omega t}$$

Based on the convolution theorem, (5.10) could be written as

$$\text{WVD}_{s,g}(t, \omega) = \int s\left(t + \frac{\tau}{2}\right)g^*\left(t - \frac{\tau}{2}\right)\exp\{-j\omega\tau\}d\tau$$

$$= \int s_1(\tau)g_1(\tau)\exp\{-j\omega\tau\}d\tau$$

$$= S_1(\omega) \otimes G_1(\omega)$$

$$= \frac{4}{2\pi}\int S(2\alpha)G^*[2\omega - 2\alpha]e^{j(4\alpha - 2\omega)t}d\alpha$$

Let $2\alpha = \omega + \Omega/2$, then,

$$\text{WVD}_{s,g}(t, \omega) = \frac{1}{2\pi}\int S\left(\omega + \frac{\Omega}{2}\right)G^*\left(\omega - \frac{\Omega}{2}\right)\exp\{j\Omega t\}d\Omega \qquad (5.15)$$

where $S(\omega)$ denotes the Fourier transform of $s(t)$ and $G(\omega)$ denotes the Fourier transform of $g(t)$. The corresponding auto-WVD is

$$\text{WVD}_s(t, \omega) = \frac{1}{2\pi}\int S\left(\omega + \frac{\Omega}{2}\right)S^*\left(\omega - \frac{\Omega}{2}\right)\exp\{j\Omega t\}d\Omega \qquad (5.16)$$

Eqs.(5.16) and (5.9) indicate that the WVD is symmetric in the time and frequency domains. Consequently, a property derived from (5.9) always has the dual property in the frequency domain.

5.2 General Properties of Wigner-Ville Distribution

In the preceding section, we introduced the concept of time-dependent spectrum and the Wigner-Ville distribution. Compared to STFT, the WVD not only has a better resolution, but also does not suffer the window effects. In what follows, we shall investigate the WVD in more detail.

Time-Shift Invariant: If the WVD of signal $s(t)$ is $\text{WVD}_s(t,\omega)$, then the WVD of the time-shifted version $s_0(t) = s(t-t_0)$ is a time-shifted WVD of $s(t)$, i.e.,

$$\text{WVD}_{s0}(t, \omega) \;=\; \text{WVD}_s(t - t_0, \omega) \qquad (5.17)$$

Frequency Modulation Invariant: If the WVD of signal $s(t)$ is $\text{WVD}_s(t,\omega)$, then the WVD of the frequency-modulated version $s_0(t) = s(t)\exp\{j\omega_0 t\}$ is a frequency-shifted WVD of $s(t)$, i.e.,

$$\text{WVD}_{s0}(t, \omega) \;=\; \text{WVD}_s(t, \omega - \omega_0) \qquad (5.18)$$

Time Marginal Condition:

$$\frac{1}{2\pi}\int \text{WVD}_s(t, \omega)d\omega \;=\; \int s\!\left(t + \frac{\tau}{2}\right)s^*\!\left(t - \frac{\tau}{2}\right)\frac{1}{2\pi}\int \exp\{-j\omega\tau\}d\omega d\tau$$

$$=\; \int s\!\left(t + \frac{\tau}{2}\right)s^*\!\left(t - \frac{\tau}{2}\right)\delta(\tau)d\tau \;=\; |s(t)|^2 \qquad (5.19)$$

Frequency Marginal Condition:

$$\int \text{WVD}_s(t, \omega)dt \;=\; \int\int s\!\left(t + \frac{\tau}{2}\right)s^*\!\left(t - \frac{\tau}{2}\right)\exp\{-j\omega\tau\}dt d\tau$$

$$=\; \int \exp\{-j\omega\tau\}\int s(t)s^*(t - \tau)dt d\tau$$

$$=\; \int \exp\{-j\omega\tau\}R(\tau)d\tau \;=\; |S(\omega)|^2$$

Because the WVD satisfies both time and frequency marginal conditions, we can readily determine from Parseval's relation that

$$\frac{1}{2\pi}\int\int \text{WVD}_s(t, \omega)d\omega dt \;=\; \frac{1}{2\pi}\int |S(\omega)|^2 d\omega \;=\; \int |s(t)|^2 dt \qquad (5.20)$$

which says that the energy contained in $\text{WVD}(t,\omega)$ is equal to the energy possessed by the original signal $s(t)$.

Example 5-2   WVD of linear chirp signal with
        Gaussian envelope

$$s(t) = \left(\frac{\alpha}{\pi}\right)^{\frac{1}{4}} \exp\left\{-\frac{\alpha}{2}t^2 + j\beta t^2\right\} \tag{5.21}$$

which is the linear chirp signal with the Gaussian envelope, the first derivative of the phase $\varphi'(t) = 2\beta t$ that linearly increases with time. The WVD is

$$\text{WVD}_s(t, \omega) = \exp\{-\alpha t^2\} \sqrt{\frac{\alpha}{\pi}} \int \exp\left\{-\frac{\alpha}{4}\tau^2\right\} \exp\{-j(\omega - 2\beta t)\tau\} d\tau$$

$$= 2\exp\left\{-\left[\alpha t^2 + \frac{1}{\alpha}(\omega - 2\beta t)^2\right]\right\} \tag{5.22}$$

which is plotted in Fig. 5-2.

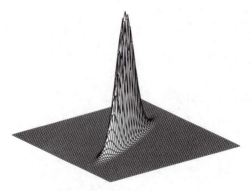

Fig. 5-2 WVD of linear chirp with Gaussian envelope is concentrated at $\omega = 2\beta t$.

If we consider (5.22) as the linear chirp energy distribution in the joint time-frequency domain, then the conditional mean frequency of (5.22) naturally is a measure of a mean instantaneous frequency, that is,

$$\langle \omega \rangle_t = \frac{\frac{1}{2\pi}\int \omega\, \text{WVD}_s(t, \omega) d\omega}{\frac{1}{2\pi}\int \text{WVD}_s(t, \omega) d\omega} = \frac{\frac{1}{2\pi}\int \omega\, \text{WVD}_s(t, \omega) d\omega}{|s(t)|^2} \tag{5.23}$$

which indicates the center of the spectrum at time instant t. Substituting (5.22) into (5.23) obtains

$$\langle\omega\rangle_t = \frac{\frac{1}{\pi}\exp\{\alpha t^2\}\int\omega\exp\left\{-\frac{1}{\alpha}(\omega-2\beta t)^2\right\}d\omega}{|s(t)|^2} = 2\beta t \qquad (5.24)$$

Note that the right side of (5.24) is exactly the first derivative of the signal's phase, that is,

$$\langle\omega\rangle_t = 2\beta t = \varphi'(t) \qquad (5.25)$$

Therefore, the first derivative of the signal's phase reflects the mean instantaneous frequency. Eq. (5.25) describes a very important property of the WVD.

As a matter of fact, the relationship (5.25) not only holds for Example Example 5–2 but also for arbitrary signals. Therefore, we name the formula (5.25) an *instantaneous frequency property*.

Instantaneous Frequency Property: Let $s(t) = A(t)\exp\{j\varphi(t)\}$, where magnitude $A(t)$ and phase $\varphi(t)$ both are real-valued functions. Then,

$$\langle\omega\rangle_t = \frac{\frac{1}{2\pi}\int\omega\,\text{WVD}_s(t,\omega)d\omega}{\frac{1}{2\pi}\int\text{WVD}_s(t,\omega)d\omega} = \frac{\frac{1}{2\pi}\int\omega\,\text{WVD}_s(t,\omega)d\omega}{|A(t)|^2} = \varphi'(t) \qquad (5.26)$$

which says that, at time instant t, the mean instantaneous frequency of WVD is equal to the mean instantaneous frequency of the analyzed signal.

PROOF

$$\frac{1}{2\pi}\int\omega\,\text{WVD}_s(t,\omega)d\omega = \left(\frac{1}{2\pi}\right)^2\int e^{j\Omega t}\int\omega S\left(\omega+\frac{\Omega}{2}\right)S^*\left(\omega-\frac{\Omega}{2}\right)d\omega d\Omega \qquad (5.27)$$

Let

$$H(\omega) = \omega S\left(\omega+\frac{\Omega}{2}\right) \qquad \text{and} \qquad G(\omega) = S\left(\omega-\frac{\Omega}{2}\right)$$

Based on derivative property (2.34) and shifting property (2.23), we have

$$h(t) = -j\frac{d}{dt}\left(s(t)\exp\left\{-j\frac{\Omega}{2}t\right\}\right) \qquad \text{and} \qquad g(t) = s(t)\exp\left\{j\frac{\Omega}{2}t\right\}$$

Applying Parseval's formula (2.39), we can rewrite (5.27) as

$$\frac{1}{2\pi}\int \omega\, WVD_s(t,\omega)d\omega = \frac{1}{2\pi}\int e^{j\Omega t}\int \left\{-j\frac{d}{da}[s(a)e^{-ja\Omega/2}]s^*(a)e^{-ja\Omega/2}\right\}dad\Omega$$

$$= \frac{1}{2\pi}\int e^{j\Omega t}\int \left\{-\frac{\Omega}{2}|s(a)|^2 e^{-ja\Omega}-je^{-ja\Omega}s^*(a)\frac{d}{da}s(t)\right\}dad\Omega$$

$$= \frac{1}{2\pi}\int e^{j\Omega t}\int -\frac{\Omega}{2}|s(a)|^2 e^{-ja\Omega}dad\Omega - \frac{j}{2\pi}\int s^*(a)\frac{d}{da}s(t)\int e^{j\Omega(t-a)}d\Omega da$$

$$= -\frac{1}{4\pi}\int \Omega e^{j\Omega t}\int |s(a)|^2 e^{-ja\Omega}dad\Omega - js^*(a)\frac{d}{da}s(t) \qquad (5.28)$$

Applying the convolution theorem, the first term becomes

$$-\frac{1}{4\pi}\int \Omega e^{j\Omega t}\int |s(a)|^2 e^{-ja\Omega}dad\Omega = -\frac{1}{4\pi}\int \Omega e^{j\Omega t}[S(\Omega)\oplus S(\Omega)]d\Omega$$

$$= \frac{1}{4\pi}\int S(b)\frac{1}{2\pi}\int \Omega S^*(\Omega-b)e^{j\Omega t}d\Omega db$$

$$= \frac{-j}{2}\int S(b)\frac{d}{dt}[s^*(t)e^{jbt}]db$$

$$= \frac{-j}{2}\int S(b)\left[e^{jbt}\frac{d}{dt}s^*(t)+jbs^*(t)e^{jbt}\right]db$$

$$= \frac{-j}{2}s(t)\frac{d}{dt}s^*(t)+\frac{j}{2}s^*(t)\frac{d}{dt}s(t) \qquad (5.29)$$

Substituting (5.29) into (5.28) obtains

$$\frac{1}{2\pi}\int \omega\, WVD_s(t,\omega)d\omega = \frac{-j}{2}\left\{s(t)\frac{d}{dt}s^*(t)+s^*(t)\frac{d}{dt}s(t)\right\} = |A(t)|^2\varphi'(t) \qquad (5.30)$$

Replacing the numerator in (5.26) by (5.30), we can readily obtain the instantaneous frequency property.

Traditionally, we use the instantaneous frequency property (5.26) to evaluate whether or not the time-dependent spectrum reflects the signal's frequency content changes. For the desired time-dependent spectrum, we usually hope that

$$\frac{\int \omega P(t,\omega)d\omega}{\int P(t,\omega)d\omega} = \varphi'(t) \qquad (5.31)$$

where $s(t) = A(t)\exp\{j\varphi(t)\}$. Note that neither the STFT spectrogram nor scalogram satisfy (5.31).

Eqs.(5.14) and (5.22) show that, at any time instant t, there is more than one frequency component. In other words, the instantaneous frequency is not a single value function of time. Signal energy spreads with respect to the mean instantaneous frequencies. The instantaneous bandwidth is not equal to zero. As a matter of fact, this not only is true for Examples Example 5–1 and Example 5–2, but also holds for any signals whose energy is finite[*]. Therefore, the next interesting question is, what is the instantaneous bandwidth? Or, how does the signal's energy spread with respect to the mean instantaneous frequency $\varphi'(t)$? Because the WVD in (5.14) and (5.22) are non-negative. We can use the concept of a conditional variance to measure the instantaneous bandwidth. For example,

$$\Delta_t^2 = \frac{\int (\omega - \langle \omega \rangle_t)^2 \, \mathrm{WVD}_s(t, \omega) d\omega}{\int \mathrm{WVD}_s(t, \omega) d\omega}$$

$$= \frac{\frac{1}{\pi} \exp\{-\alpha t^2\} \int (\omega - \langle \omega \rangle_t)^2 \exp\left\{-\frac{1}{\alpha}(\omega - 2\beta t)^2\right\} d\omega}{|s(t)|^2} = \frac{\alpha}{2}$$

which indicates that the energy spread is independent of time.

Unfortunately, except in a few cases, in general the WVD could go to negative. Consequently, we cannot simply apply the conventional variance concept to measure the instantaneous bandwidth. When a WVD is negative, its conditional variance (or signal's energy spread) may become negative too, which obviously does not make sense. In this case, special care has to be taken when the instantaneous bandwidth is evaluated.

Group Delay Property: Assume that the Fourier transform of signal $s(t)$ is $S(\omega) = B(\omega)\exp\{j\psi(\omega)\}$. Then, the first derivative of the phase $\psi'(\omega)$ is called the *group delay*. For Wigner-Ville distribution, we have

$$\frac{\int t \, \mathrm{WVD}_s(t, \omega) dt}{\int \mathrm{WVD}_s(t, \omega) dt} = \frac{\int t \, \mathrm{WVD}_s(t, \omega) dt}{|S(\omega)|^2} = -2\pi\psi'(\omega) \tag{5.32}$$

which says that the conditional mean time of WVD is equal to the group delay. We leave the proof for the reader to practice.

It is important to note that most useful properties of the WVD are determined by *averaging* the WVD. For instance, the time marginal condition is obtained by averaging the WVD over the frequency. The instantaneous frequency property (5.26) is the conditional frequency average. This suggests that

---

[*] As introduced in Section 2.4, the frequency bandwidth is determined by variations of the phase as well as magnitude. Zero instantaneous bandwidth requires no magnitude variations. In other words, the magnitude has to be constant. In this case, the resulting signal's energy is unbounded, such as $\exp\{j\omega_0 t\}$ and $\exp\{j\alpha t^2\}$.

the properties that are useful to signal processing will be mainly determined by the smooth portions. Because the average of the high oscillation portion is small, the highly oscillated portion will have limited influence on the useful properties. This is a very important observation which leads to an improved time-dependent spectrum, time-frequency distribution series. We shall discuss this subject a great deal in Chapter 7.

Although many other time-dependent spectrum schemes claim to preserve all useful properties of the WVD, none of them has such desired representation for the signal with the Gaussian envelope as does the WVD in (5.14) and (5.22). They may satisfy the marginal conditions and instantaneous frequency property, but they have negative values and are not as concentrated as (5.14) and (5.22).

Finally, as in the case of the short-time Fourier transformation, not all time-frequency functions $P(t,\omega)$ can be the Wigner-Ville distribution. For example,

$$P(t, \omega) = \begin{cases} 1 & |t| \leq t_0 \ and \ |\omega| \leq \omega_0 \\ 0 & \text{otherwise} \end{cases} \tag{5.33}$$

will not be a valid Wigner-Ville distribution, because no signal can be time and frequency limited simultaneously.

5.3   WVD of Sum of Multiple Signals

As introduced in the preceding sections, the Wigner-Ville distribution not only possesses many useful properties, but also has better resolution than the STFT spectrogram. Although the WVD has existed for a long time, its applications are very limited. One main deficiency of the WVD is the so-called cross-term interference. For $s(t) = s_1(t) + s_2(t)$, the WVD is

$$\text{WVD}_s(t, \omega) = \text{WVD}_{s1}(t, \omega) + \text{WVD}_{s2}(t, \omega) + 2Re\{ \text{WVD}_{s1,s2}(t, \omega)\} \tag{5.34}$$

which shows that the WVD of the sum of two signals is not the sum of their respective WVDs. In addition to two auto-terms, (5.34) contains one cross-term $\text{WVD}_{s1,s2}(t,\omega)$. Because the cross-term usually oscillates and its magnitude is twice as large as that of the auto-terms, it often obscures the useful time-dependent spectrum patterns. How to reduce the cross-term interference without destroying the useful properties of the WVD has been very important to time-frequency analysis. This will also be the focus of the rest of Part 2. Before introducing various joint time-frequency representation schemes, let's look at a couple of examples to get a better idea about the cross-term interference.

Example 5–3   WVD of complex sinusoidal signals

When $s(t)$ is a single complex sinusoidal signal, for instance, $s(t) = exp(j\omega_0 t)$, we have

$$\text{WVD}_s(t, \omega) = \int \exp\left\{j\omega_0\left(t + \frac{\tau}{2} - t + \frac{\tau}{2}\right)\right\}\exp\{-j\omega\tau\}d\tau = 2\pi\delta(\omega - \omega_0) \qquad (5.35)$$

which shows that the WVD is concentrated at frequency ω. This is exactly what we expected. When $s(t)$ consists of the sum of two complex sinusoidal signals, such as $s(t) = \exp(j\omega_1 t) + \exp(j\omega_2 t)$, the conventional power spectrum is

$$\text{PS}(\omega) = 2\pi\delta(\omega_1) + 2\pi\delta(\omega_2) \qquad (5.36)$$

The WVD is

$$\text{WVD}_s(t, \omega) = 2\pi \sum_{i=1}^{2} \delta(\omega - \omega_i) + 4\pi\delta(\omega - \omega_\mu)\cos\{\omega_d t\} \qquad (5.37)$$

where ω_μ and ω_d denote the geometric center and the distance between two complex sinusoidal functions in the frequency domain, i.e.,

$$\omega_\mu = \frac{\omega_1 + \omega_2}{2} \qquad \omega_d = \omega_1 - \omega_2$$

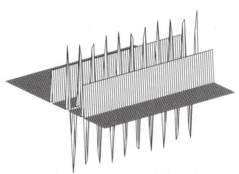

Fig. 5–3 The WVD of two sinusoidal signals contains two desired auto-terms at frequencies ω_1 and ω_2, and a large cross-term midway between those auto-terms.

Formula (5.37) is plotted in Fig. 5–3. In addition to the two desired auto-terms at frequencies ω_1 and ω_2, we get a large cross-term at ω_μ, midway between those two auto-terms. Unlike the two auto-terms that are non-negative, the cross-term oscillates at frequency ω_d, the distance between the two individual components in the frequency domain. Although the average of the cross-term is equal to zero, i.e.,

$$\int 4\pi\delta(\omega - \omega_\mu)\cos\{\omega_d t\}dt = 0 \qquad \text{for } \omega_d \neq 0 \qquad (5.38)$$

the magnitude of the cross-term is twice as large as the auto-terms! Because the conventional power spectrum in (5.37) indicates that there is no signal at ω_μ, and the energy contained in the cross-term is zero, the cross-term is commonly considered as interference.

From formula (5.34), each pair of auto-terms creates one cross-term. For N individual components, the total number of cross-terms is $N(N-1)/2$. In the simple case, such as in Example 5–4, we can easily identify the cross-term interference. For real signals, the pattern of the cross-terms, which usually overlap with auto-terms, could be more complicated. Consequently, the desired time-dependent spectrum could be deceiving and confusing.

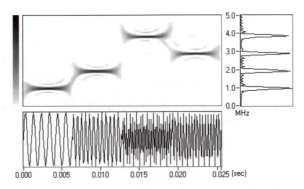

Fig. 5–4 The bottom plot is a time waveform that contains four frequency tones. The right plot is the traditional power spectrum. The middle one is the joint time-frequency representation computed by *time-frequency distribution series*.

Fig. 5–4 illustrates the sum of four frequency tones. The bottom plot is time waveforms. The right plot is the traditional power spectrum. The middle plot is the desired time-dependent spectrum computed by the *time-frequency distribution series* that we will discuss in Chapter 7. The conventional power spectrum indicates that there are four different frequency tones, but it is not clear when those different frequency tones occur. The time-dependent power spectrum not only shows four frequency tones, but also tells when they take place.

Fig. 5–5 and Fig. 5–6 illustrate the WVD of the same frequency hopper signal. Because of the presence of the cross-term interference, the useful time-dependent pattern is completely destroyed. It is those undesired terms that obscure the application of the WVD, even though the WVD possesses many important properties for signal analysis.

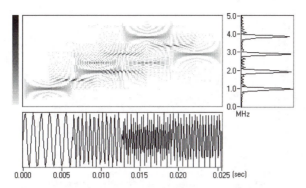

Fig. 5–5 Frequency hopper signal computed by WVD.

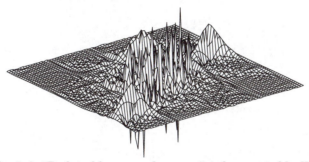

Fig. 5–6 3D plot of frequency hopper signal computed by WVD.

Fig. 5–7 and Fig. 5–8 demonstrate seismic signals. The power spectrum clearly indicates two very strong frequency components. Because of the presence of the interference, however, those two prominent frequency tones are completely destroyed in the WVD (see Fig. 5–8).

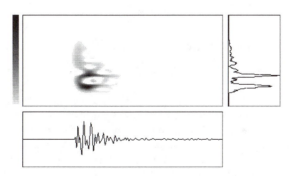

Fig. 5–7 Seismic signal computed by time-frequency distribution series. (Data courtesy of Naval Research Laboratory, Washington, D.C.)

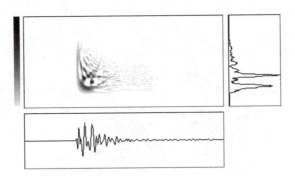

Fig. 5–8 Seismic signal computed by WVD.

Example 5-4   WVD of two Gaussian functions

$$s(t) = \sum_{i=1}^{2} \left(\frac{\alpha}{\pi}\right)^{\frac{1}{4}} \exp\left\{-\frac{\alpha}{2}(t-t_i)^2 + j\omega_i t\right\} \qquad (5.39)$$

which contains two frequency-modulated Gaussian signals. One is concentrated in (t_1, ω_1). The other is centered in (t_2, ω_2).   Then,

$$\text{WVD}_s(t, \omega) = 2 \sum_{i=1}^{2} \exp\left\{-\alpha(t-t_i)^2 - \frac{1}{\alpha}(\omega - \omega_i)^2\right\} \qquad (5.40)$$

$$+ 4\exp\left\{-\alpha(t-t_\mu)^2 - \frac{1}{\alpha}(\omega - \omega_\mu)^2\right\} \cos[(\omega - \omega_\mu)t_d + \omega_d(t-t_\mu) + \omega_d t_\mu]$$

where t_μ denotes the geometric time mean between two individual Gaussian signals, that is,

$$t_\mu = \frac{t_1 + t_2}{2}$$

t_d is the distance between two individual functions in time domain, i.e.,

$$t_d = t_1 - t_2$$

The summation in (5.40) corresponds to the auto-terms, which are non-negative. The last term represents the cross-term at (t_μ, ω_μ), midway between the two auto-terms, as shown in Fig. 5–9. The cross-term oscillates in both time and frequency directions. The rate of oscillation is proportional to t_d and ω_d, the distance between two auto-terms. Both auto-terms and cross-term have the same two-dimensional Gaussian envelope that is concentrated and symmetrical in joint time-frequency domain.

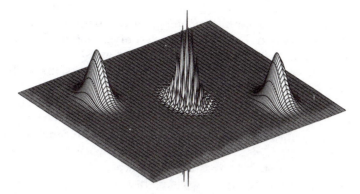

Fig. 5–9 The cross-term is at (t_μ, ω_μ), midway between the two auto-terms. It oscillates in both time and frequency directions.

Example Example 5–4 demonstrates the mechanism of the cross-term. The cross-term in fact reflects the correlation of the corresponding pair of auto-terms. Its location and the rate of oscillation are determined by time and frequency centers of the auto-terms. If we know the positions of the auto-terms then we can precisely identify the corresponding cross-term. In the case of two Gaussian functions, both auto-terms and cross-term are symmetrical and concentrated in the joint time-frequency domain. Now, the interesting problem is how the auto-term and cross-term affect the useful properties. To answer this question, let's compute some useful properties, such as the time marginal condition.

Inserting (5.40) into (5.19) yields

$$|s(t)|^2 = \sqrt{\frac{\alpha}{\pi}} \sum_{i=1}^{2} \exp\{-\alpha(t-t_i)^2\} + 2\sqrt{\frac{\alpha}{\pi}} \exp\left\{-\frac{\alpha}{4} t_d^2\right\} \exp\{-\alpha(t-t_\mu)^2\} \cos\{\omega_d t\}$$

$$= \sqrt{\frac{\alpha}{\pi}} \sum_{i=1}^{2} \exp\{-\alpha(t-t_i)^2\} + A(t_d^2) \exp\{-\alpha(t-t_\mu)^2\} \cos(\omega_d t) \qquad (5.41)$$

The summation is the integration of the auto-terms, which are non-negative. The second term is caused by the cross-term, which oscillates. The rate of oscillation ω_d is proportional to the distance between two auto-terms in frequency domain. The magnitude $A(t_d^2)$ exponentially decays as t_d, the distance between two auto-terms in the time domain, increases. In another words, the farther apart the auto-terms are, the less energy the cross-term contains. Similar observations also hold for the other properties. For example, the reader can verify that only the cross-terms that are created by close auto-terms have significant influence on the mean instantaneous frequency.

Traditionally, the cross-term has been used to describe something undesired in the WVD, such as those shown in Example 5–3 and Example 5–4. Consequently, they should all be removed. Actually, this thinking is not entirely correct. First, unless for very simple cases, such as two sinusoidal signals in Example 5–3 and two Gaussian functions in Example 5–4, the cross-term in general is not well defined. As we know, the signal can be broken into parts in an infinite number of ways, and the different decomposition will lead to different cross-terms. Second, the cross-term is not always a ghost. As shown in Example 5–4, when the cross-term is created by two Gaussian functions whose time and frequency centers are far apart, the cross-term highly oscillates, which has limited influence on the time marginal conditions as well as other useful properties. In this case, the cross-term is indeed not important and thereby could be removed. When the cross-term is caused by two Gaussian functions whose time and frequency centers are closer, the cross-term will oscillate less and therefore has a larger average. If such a cross-term is discarded, the resulting presentation will leave significant signal energy out. In this case, we can expect to lose much of the time marginal condition as well as other properties. Thus, simply discarding all cross-terms is not a right approach.

At this point, the natural question is, what parts of the WVD are unwanted? How can we get a better joint time-frequency representation? Before fully addressing this question, let's further investigate some other aspects of the WVD.

5.4   Wigner-Ville Distribution, STFT Spectrogram, and Scalogram

In the beginning of this chapter, we mentioned that both STFT spectrogram and scalogram can be used to describe a signal's energy changes in joint time-frequency domain. Then, how are STFT spectrogram and scalogram related to the Wigner-Ville distribution? In what follows, we shall apply elementary multivariable calculus to introduce some important relations.

For an arbitrary signal $s(t)$,

$$\mathrm{PS}_s(t, \omega) = \left| \mathrm{STFT}_s(t, \omega) \right|^2 = \iint \mathrm{WVD}_s(x, y) \mathrm{WVD}_\gamma(t - x, \omega - y) dx dy \qquad (5.42)$$

where $\mathrm{WVD}_s(t,\omega)$ and $\mathrm{WVD}_\gamma(t,\omega)$ denote the WVD of the analyzed signal $s(t)$ and the analysis function $\gamma(t)$, respectively. Formula (5.42) is a typical 2D convolution, which says that the STFT spectrogram in fact is the convolution of the Wigner-Ville distributions of the signal $s(t)$ and analysis function $\gamma(t)$.

PROOF

Let's expand the right side of (5.42).

$$\iiiint s\left(x+\frac{\mu}{2}\right)s^*\left(x-\frac{\mu}{2}\right)\gamma\left(t-x+\frac{v}{2}\right)\gamma^*\left(t-x-\frac{v}{2}\right)\exp\{-jy(\mu-v)-jv\omega\}d\mu dv dx dy$$

$$=\iiint s\left(x+\frac{\mu}{2}\right)s^*\left(x-\frac{\mu}{2}\right)\gamma\left(t-x+\frac{v}{2}\right)\gamma^*\left(t-x-\frac{v}{2}\right)\exp\{-jv\omega\}\delta(\mu-v)d\mu dv dx$$

$$=\iint s\left(x+\frac{v}{2}\right)s^*\left(x-\frac{v}{2}\right)\gamma\left(t-x+\frac{v}{2}\right)\gamma^*\left(t-x-\frac{v}{2}\right)\exp\{-jv\omega\}dv dx \qquad (5.43)$$

Let

$$a = x+\frac{v}{2} \qquad and \qquad b = x-\frac{v}{2}$$

Then

$$x = \frac{a+b}{2} \qquad v = a-b$$

The Jacobian determinant is

$$J = \begin{vmatrix} \dfrac{\partial x}{\partial a} & \dfrac{\partial x}{\partial b} \\ \dfrac{\partial v}{\partial a} & \dfrac{\partial v}{\partial b} \end{vmatrix} = \begin{bmatrix} 0.5 & 0.5 \\ 1 & -1 \end{bmatrix} = -1 \qquad (5.44)$$

Because

$$dx dv = |J| da db = da db \qquad (5.45)$$

formula (5.44) becomes

$$\iint s(a)s^*(b)\gamma(t-b)\gamma^*(t-a)\exp\{-j(a-b)\omega\}da db$$

$$= \int s(a)\gamma^*(t-a)\exp\{-j\omega a\}da\int s^*(b)\gamma(t-b)\exp\{j\omega b\}db$$

$$= |STFT_s(t,\omega)|^2$$

When the Wigner-Ville distribution of the analysis function $WVD_\gamma(t,\omega)$ is of lowpass, as in the case of most applications, then the STFT spectrogram is a smoothed WVD. This implies that the resolution of the STFT spectrogram is inferior to that of the WVD. Fig. 5–10 depicts the STFT spectrogram for the frequency hopped signal. In this example, the analysis function is a Gaussian, whose WVD is 2D lowpass. Compared to the WVD in Fig. 5–5, the STFT spectrogram has a poor resolution, but there is no interference as that which appears in the WVD.

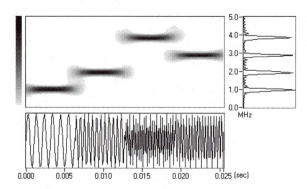

Fig. 5–10 Compared to the WVD in Fig. 5–5, the STFT has lower resolutions but it does not have unwanted interference.

It is interesting to note that the scalogram, the square of the wavelet transform, can be written in a very similar way [159], e.g.,

$$\text{SCAL}(a, b) = |\text{CWT}(a, b)|^2 = \iint \text{WVD}_s(x, y)\text{WVD}_\psi\left(\frac{x-b}{a}, ay\right)dxdy \qquad (5.46)$$

where $\text{WVD}_s(t,\omega)$ and $\text{WVD}_\psi(t,\omega)$ denote the WVD of the analyzed signal $s(t)$ and the mother wavelet function $\psi(t)$, respectively. The operation which is involved in (5.46) is now an *affine correlation*.

PROOF

Let's expand the right side of (5.46), i.e.,

$$\iiiint s\left(x + \frac{m}{2}\right)s^*\left(x - \frac{m}{2}\right)\psi\left(\frac{x-b}{a} + \frac{n}{2}\right)\psi^*\left(\frac{x-b}{a} - \frac{n}{2}\right)$$

$$\times \exp\{-j(m + an)y\}dxdydmdn$$

$$= \iiint s\left(x + \frac{m}{2}\right)s^*\left(x - \frac{m}{2}\right)\psi\left(\frac{x-b}{a} + \frac{n}{2}\right)\psi^*\left(\frac{x-b}{a} - \frac{n}{2}\right)\delta(m + an)dxdmdn$$

$$= \iint s\left(x + \frac{an}{2}\right)s^*\left(x - \frac{an}{2}\right)\psi\left(\frac{x-b}{a} + \frac{n}{2}\right)\psi^*\left(\frac{x-b}{a} - \frac{n}{2}\right)dxdn \qquad (5.47)$$

Let $u = x+(an/2)$ and $v = x-(an/2)$. Then

$$x = \frac{u + v}{2} \qquad and \qquad n = \frac{u - v}{a}$$

The Jacobian determinant is

$$J = \begin{vmatrix} \dfrac{\partial x}{\partial u} & \dfrac{\partial x}{\partial u} \\[2mm] \dfrac{\partial n}{\partial u} & \dfrac{\partial n}{\partial v} \end{vmatrix} = \begin{bmatrix} 0.5 & 0.5 \\[1mm] \dfrac{1}{a} & -\dfrac{1}{a} \end{bmatrix} = -\dfrac{1}{a} \tag{5.48}$$

Because

$$xdn = |J|dudv = \frac{dud}{a} \tag{5.49}$$

the formula (5.47) becomes

$$\iint s(u)\psi\left(\frac{u-b}{a}\right)s^*(v)\psi^*\left(\frac{v-b}{a}\right)\frac{dudv}{a} = |\mathrm{CWT}(a,b)|^2$$

5.5 Smoothed WVD and WVD of Analytical Signals

In the preceding section, we showed that the WVD of the sum of multicomponents is the linear combination of auto- and cross-terms. The auto-terms, in general, are relatively smooth, whereas the cross-terms are strongly oscillated. Therefore, a natural way of lowering cross-term interference is to apply a lowpass filter $H(t,\omega)$ to the WVD, i.e.,

$$\mathrm{SWVD}_s(t,\omega) = \iint \mathrm{WVD}_s(x,y)H(t-x,\omega-y)dxdy \tag{5.50}$$

Because the lowpass filter performs a smoothing operation, (5.50) is called the smoothed Wigner-Ville distribution (SWVD). Usually, the lowpass filtering can substantially suppress the cross-terms. On the other hand, however, smoothing will reduce the resolution. A trade-off exists between the degree of smoothing and the resolution.

If $H(t,\omega)$ is the WVD of a function $\gamma(t)$, then (5.50) becomes the STFT spectrogram, as shown previously. In this case, the SWVD manifestly is non-negative. However, the STFT spectrogram does not preserve time marginal, frequency marginal, instantaneous frequency, and many other useful properties that are possessed by the WVD. Usually, the SWVD improves the cross-term interference at the cost of lower resolution and the loss of other useful properties.

Note that formula (5.50) applies for all bilinear transformations. Any bilinear transform $C_s(t,\omega)$ can be expressed as the convolution of $\mathrm{WVD}_s(t,\omega)$ and some 2D filter $H(t,\omega)$ [32], i.e.,

$$C_s(t,\omega) = \iint \mathrm{WVD}_s(x,y)H(t-x,\omega-y)dxdy \tag{5.51}$$

However, smoothing only makes sense when the 2D filter $H(t,\omega)$ is lowpass.

Conversely, we can also write the Wigner-Ville distribution as the convolution of any other bilinear transform $C_s(t,\omega)$ with 2D filter $G(t,\omega)$, i.e.,

$$\mathrm{WVD}_s(t, \omega) = \iint C_s(x, y)G(t - x, \omega - y)dxdy \tag{5.52}$$

unless $G(t,\omega)$ is a 2D lowpass filter, otherwise $\mathrm{WVD}_s(t,\omega)$ will not be a smoothed version of $C_s(t,\omega)$. For example, we usually do not say that a *Wigner-Ville distribution* is a smoothed Choi-Williams distribution, because $G(t,\omega)$ is not the lowpass in this case.

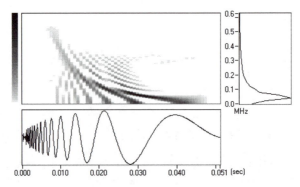

Fig. 5–11 WVD of Doppler signal (real-valued).

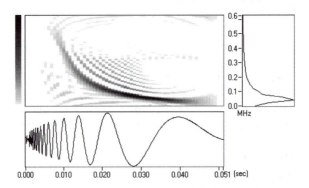

Fig. 5–12 WVD of Doppler signal (analytic signal).

The signals that we dealt with under normal circumstances are real-valued. A direct consequence of the realness is that the signal's spectrum is always symmetric, that is

$$S(\omega) = S^*(-\omega) \qquad \text{or} \qquad |S(\omega)|^2 = |S(-\omega)|^2 \tag{5.53}$$

In this case, only one half of the spectrum contains information, while the other half is redundant. To remove the redundancy, people usually convert the real signal to the so-called analytical signal. If $S(\omega)$ denotes the Fourier transform of the real-valued signal $s(t)$, then the analytical signal is the inverse Fourier transform of $S_a(\omega)$ given by

$$S_a(\omega) = \begin{cases} 2S(\omega) & \omega > 0 \\ S(\omega) & \omega = 0 \\ 0 & \omega < 0 \end{cases} \qquad (5.54)$$

For a real-valued signal, the negative frequency components not only introduce the redundancy, but also create cross-terms. To reduce the cross-term interference, Boashash [14] suggested using the analytical signal's WVD. Because the analytical function is a halfhand function, the resulting $\text{WVD}_a(t,\omega)$ avoids all cross-terms that are associated with the negative frequency components.

Fig. 5–11 and Fig. 5–12 illustrate the WVD of real-valued Doppler signal and the WVD_a of the corresponding analytical signal. Obviously, interference is much less in Fig. 5–12 than that in Fig. 5–11.

However, it has to be kept in mind that the analytical function differs from the original signal in several ways (see Cohen [32]). For instance, the analytical function of a time-limited real-valued signal $s(t)$, where $s(t) = 0$ for $t < t_1$ and $t > t_2$, is manifestly no longer time limited, because the analytical function is band limited. Therefore, we should be very careful when the analytical function is used. Although the analytical function has the same positive power spectrum as that of a corresponding real-valued signal, its instantaneous properties may substantially differ from that of the original signal.

The relation between the WVD_a and WVD can best be determined by using (5.16). For example,

$$\text{WVD}_a(t, \omega) = \frac{1}{2\pi} \int_{-\infty}^{\infty} S_a\left(\omega + \frac{\Omega}{2}\right) S_a^*\left(\omega - \frac{\Omega}{2}\right) \exp\{j\Omega t\} d\Omega$$

$$= \frac{1}{2\pi} \int_{-2\omega}^{2\omega} S\left(\omega + \frac{\Omega}{2}\right) S^*\left(\omega - \frac{\Omega}{2}\right) \exp\{j\Omega t\} d\Omega$$

which is equivalent to

$$\text{WVD}_a(t, \omega) = \frac{1}{2\pi} \int_{-\infty}^{\infty} H(\Omega) S\left(\omega + \frac{\Omega}{2}\right) S^*\left(\omega - \frac{\Omega}{2}\right) \exp\{j\Omega t\} d\Omega \qquad (5.55)$$

where $H(\Omega)$ is an ideal lowpass filter with cut-off frequency 2ω. Based on the convolution theorem, we can rewrite equation (5.55) as

$$\text{WVD}_a(t, \omega) = 2 \int_{-\infty}^{\infty} \frac{\sin(2\omega\tau)}{\tau} \text{WVD}_s(t - \tau, \omega) d\tau \qquad (5.56)$$

This result is to convolve the WVD with an ideal frequency-dependent lowpass filter $\sin(2\omega t)/t$. The spread of the low-frequency portion of the WVD is much wider than that of the high-frequency portion.

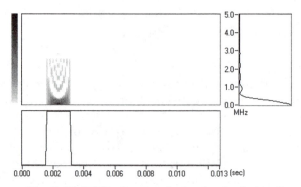

Fig. 5–13 WVD of rectangle (real-valued signal).

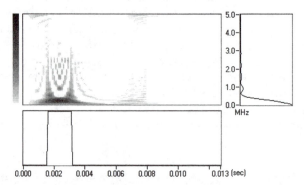

Fig. 5–14 The analytical function of the time-limited signal $s(t)$ is not time limited. The low frequency portion of the WVD_a significantly spreads out compared to the WVD for the real signal in Fig. 5–13.

Fig. 5–13 and Fig. 5–14 illustrate the WVD_s and WVD_a of the *rectangle window function*. Fig. 5–13 shows that the $\text{WVD}_s(t,\omega) = 0$ whenever $s(t) = 0$. $\text{WVD}_s(t,\omega)$ satisfies the time marginal condition. However, it is not the case for the WVD_a in Fig. 5–14. Compared to the WVD_s, the low-frequency portion of the WVD_a is significantly altered. Because the WVD_a of the analytical function is smoothed in the time domain, as shown in (5.56), all time domain properties of WVD_s, such as time marginal condition and instantaneous frequency property, are affected. The WVD_a reduces the cross-term interference at the cost of losing some useful properties.

5.6 Discrete Wigner-Ville Distribution

The continuous-time Wigner-Ville distribution introduced in the previous sections is of great value in analyzing and gaining insight into the properties of continuous-time signals. Because the majority of signals that we deal with are discrete-time signals, in the present section we shall address the subject of discrete Wigner-Ville distribution. The importance of developing the discrete counterpart is due to the increasing use and capabilities of digital computers and the development of design methods for sampled-data systems. The great flexibility of the digital computer has spurred experimentation with the design of increasingly sophisticated discrete-time systems for which no apparent practical implementation using analog equipment exists.

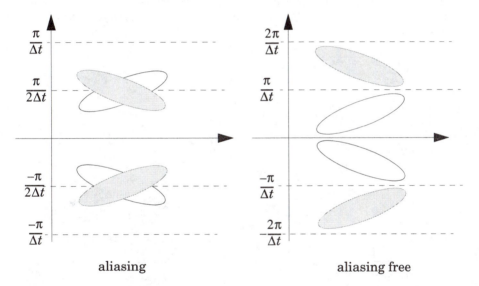

Fig. 5–15 DWVD with aliasing (left) and aliasing-free DWVD (right).

By letting $u = \tau/2$ in (5.9), the WVD becomes

$$\text{WVD}_s(t, \omega) = 2\int s(t + u)s^*(t - u)\exp\{-j2\omega u\}\,du \tag{5.57}$$

Assume the interval of the integration (5.57) is Δ. Applying the *Trapezoidal rule*, we have an approximation of the integral by

$$\overline{\text{WVD}}_s(t, \omega) = 2\Delta\sum_n s(t + n\Delta)s^*(t - n\Delta)\exp\{-j2\omega n\Delta\} \tag{5.58}$$

If the signal $s(t)$ is sampled in every Δt second, $\Delta t = \Delta$, then we obtain the discrete-time Wigner-Ville distribution as

$$\overline{\text{WVD}}_s[m\Delta t, \omega] = 2\Delta t \sum_n s[(m+n)\Delta t] s^*[(m-n)\Delta t] \exp\{-j2\omega n\Delta t\} \qquad (5.59)$$

Obviously,

$$\overline{\text{WVD}}_s[m\Delta t, \omega + \frac{\pi}{\Delta t}) = \overline{\text{WVD}}_s[m\Delta t, \omega) \qquad (5.60)$$

Therefore, the period of $\overline{\text{WVD}}_s[m\Delta t, \omega)$ is $\pi/\Delta t$ rather than $2\pi/\Delta t$ required by Shannon sampling theory. The formula (5.60) implies that the highest frequency component in (5.60) must be less than or equal to $\pi/(2\Delta t)$. If the signal bandwidth is larger than $\pi/(2\Delta t)$, then aliasing will occur (see Fig. 5–15). In order to obtain an aliasing-free discrete-time Wigner-Ville distribution, we have to double the sampling rate.

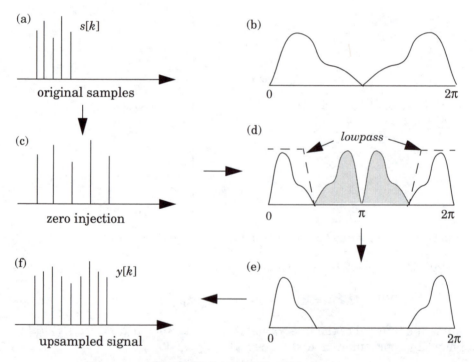

Fig. 5–16 Interpolation filtering.

The simplest way of double sampling is to apply the interpolation filter (Example 2–5). Fig. 5–16 depicts the procedure for interpolation filtering. To do an interpolation filter, first insert a zero between each sample. Then apply a lowpass filter to remove the image (shaded area in power spectrum). Such an upsampling process can be described by

$$y[m] = \sum_{n=0}^{N} s[n]\gamma[m-2n] \qquad (5.61)$$

where $\gamma[m]$ is a half-band lowpass filter. $y[m]$ denotes an upsampled sequence. If the original sample interval is Δt second, then the interval of the interpolated samples is $0.5\Delta t$ second. Applying this result into (5.59), we have

$$\overline{\text{WVD}}_s[m\frac{\Delta t}{2}, \omega) = 2\frac{\Delta t}{2}\sum_n y\Big[[m+n]\frac{\Delta t}{2}\Big]y^*\Big[[m-n]\frac{\Delta t}{2}\Big]\exp\Big\{-j2\omega n\frac{\Delta t}{2}\Big\} \qquad (5.62)$$

The period of (5.62) becomes $2\pi/\Delta t$, which is exactly what we anticipate.

Let the normalized frequency $\theta = \omega\Delta t/2$. Without loss of generality, we further assume that $\Delta t = 2$, then formula (5.62) reduces to

$$\overline{\text{WVD}}_s[m, \theta) = 2\sum_{n=-\infty}^{\infty} y[m+n]y^*[m-n]\exp\{-j2\theta n\} \qquad (5.63)$$

Formula (5.63) requires an evaluation from minus infinite to plus infinite, which is impossible in real applications. To overcome this problem, we impose a running window $w[n]$, such as

$$\overline{\text{WVD}}_s[m, \theta) = 2\sum_{n=-\infty}^{\infty} w[n]y[m+n]y^*[m-n]\exp\{-j2\theta n\} \qquad (5.64)$$

which is called the *pseudo Wigner-Ville distribution*. For the sake of simplicity, let $w[n]$ be a rectangular window[*] with length $4L+1$, i.e.,

$$w[n] = \begin{cases} 1 & |n| < 2L \\ 0 & \text{otherwise} \end{cases} \qquad (5.65)$$

---

[*] The results obtained later in fact apply for any windows which are symmetric with respect to zero, e.g., $w[n] = w[-n]$.

Substituting it into (5.64) yields

$$\overline{\mathrm{WVD}}_s[m, \theta] = 2 \sum_{n = -(2L-1)}^{2L-1} y[m+n]y^*[m-n]\exp\{-j2\theta n\}$$

$$= 2 \sum_{n = -(2L-1)}^{0} y[m+n]y^*[m-n]\exp\{-j2\theta n\}$$

$$+ 2 \sum_{n=0}^{2L-1} y[m+n]y^*[m-n]\exp\{-j2\theta n\} - 2y[m]y^*[m]$$

$$= 4Re\left\{ \sum_{n=0}^{2L-1} y[m+n]y^*[m-n]\exp\{-j2\theta n\} \right\} - 2y[m]y^*[m] \qquad (5.66)$$

If we further digitize the frequency index θ, then (5.66) becomes

$$\mathrm{DWVD}_s[m, k] = 4Re\left\{ \sum_{n=0}^{2L-1} y[m+n]y^*[m-n]\exp\left\{ -j\frac{4\pi k n}{2L} \right\} \right\} - 2y[m]y^*[m] \quad (5.67)$$

for $0 \le k < 2L$. The formula (5.67) is called the *discrete Wigner-Ville distribution* (DWDV). Note that, from (5.67)

$$\mathrm{DWVD}[m, k] = \mathrm{DWVD}[m, k+iL] \qquad \text{for } i = 0, \pm1, \pm2... \qquad (5.68)$$

If we use the standard $2L$-point FFT algorithm to compute the summation in (5.67), half the output is redundant. Usually, we assume that

$$y[m+n]y^*[m-n] = 0 \qquad \text{for } |n| > L \qquad (5.69)$$

then the formula (5.67) reduces to

$$\mathrm{DWVD}_s[m, k] = 4Re\left\{ \sum_{n=0}^{L-1} y[m+n]y^*[m-n]\exp\left\{ -j\frac{2\pi k n}{L} \right\} \right\} - 2y[m]y^*[m] \quad (5.70)$$

Now, the summation becomes L-point FFT. There is no redundancy. The frequency range of DWVD in (5.70) is $0 \le k < L$. The highest frequency without aliasing is $\pi/\Delta t$. Because the formula (5.70) is computationally efficient, it is often used to estimate the discrete Wigner-Ville distribution.

Many algorithm tricks have been proposed to overcome the aliasing problem. However, no matter what kinds of tricks are used, the requirement is usually the same: we must oversample either in the time domain or in frequency domain. Although there are some variations among the algorithms, the improvements in computational complicity as well as in memory usage are marginal. There are two advantages of the method introduced in this section. First, it is closer to the continuous-time WVD. Consequently, many properties of the continuous-time WVD are carried over to its discrete counterpart. Second, the interpolation filter is relatively easier to implement with existing hardware techniques.

It is worth noting that when the analytical signal is employed, the aliasing problem is automatically avoided because it is a half-band signal. The negative frequency components of the analyzed signal are zero. However, this method only works for real-valued signals. For complex signals, such as those encountered in the radar applications, the spectra are not symmetric and thereby we can't neglect the negative frequencies. In this case, the aliasing problem still exists unless we apply interpolation filtering or other preprocessing techniques to the signal before computing the discrete Wigner-Ville distribution.

Summary

Based upon the traditional power spectrum, in this chapter we introduced a general formula of the time-dependent spectrum. In particular, we discussed the Wigner-Ville distribution. The reason that makes the WVD so special is that the WVD better characterizes a signal's frequency changes than any other schemes, such as the STFT spectrogram and scalogram, which are popular in the area of signal processing. The WVD possesses many useful properties for signal processing. Many of them are related to the average of the WVD. Because the average of highly oscillated portions presumably possess relatively small averages, the useful properties of the WVD are mainly determined by its smooth portions.

One of the major deficiencies of the WVD is the cross-term interference. To reduce the unwanted terms, we discussed two simple methods, the smoothed WVD and the WVD of the analytical function. Although they are simple and quite effective in many applications, they all reduce the cross-terms at the cost of losing useful properties. The topic of cross-term suppression with minimum cost will be the main focus in the rest of this book.

Finally, we introduced a feasible numerical implementation of the Wigner-Ville distribution.

CHAPTER 6

Cohen's Class

*I*n Chapter 5, we introduced the Wigner-Ville distribution. In addition to the Wigner-Ville distribution, there are dozens of other bilinear joint time-frequency representations that have been developed over the last fifty years (see [23], [26], [139], [140], [156], [197], [30], and [73]) It is interesting to note that all those bilinear representations can be written in a general form that was introduced by Cohen [26]. The discovery of the general form of bilinear joint time-frequency representations facilitates us with the design of the desired joint time-frequency representations.

Because an easy way of studying Cohen's class is from the ambiguity domain, this chapter starts with a brief review of the symmetric ambiguity function. The general properties of the ambiguity function have been thoroughly studied in the context of radar and sonar. The discussion in this chapter will be focused only on those aspects that are of importance for a better understanding of Cohen's class. In Section 6.2, we discuss the relationship between the kernel function and the corresponding bilinear representations. Finally, in Section 6.3, we study a few well-known members of Cohen's class in brief, which include the Choi-Williams distribution, the cone-shape distribution, and signal-dependent time-frequency distribution.

6.1   Ambiguity Function

In the preceding chapter, we generalized the traditional power spectrum into the time-dependent spectrum as Eq.(5.3):

$$P(t, \omega) = \int R(t, \tau) \exp\{-j\omega\tau\} d\tau$$

If the time-dependent auto-correlation function is chosen as Eq.(5.8):

$$R(t, \tau) = s\left(t + \frac{\tau}{2}\right) s^*\left(t - \frac{\tau}{2}\right)$$

then the resulting time-dependent power spectrum is the Wigner-Ville distribution (WVD) (see Eq.(5.9)):

$$\text{WVD}_s(t, \omega) = \int s\left(t + \frac{\tau}{2}\right) s^*\left(t - \frac{\tau}{2}\right) \exp\{-j\omega\tau\} d\tau$$

If we take the Fourier transform with respect to the variable t instead of τ in (5.8), then we obtain another popular joint time-frequency representation called the *symmetric ambiguity function* (AF), i.e.,

$$\text{AF}_s(\vartheta, \tau) = \int s\left(t + \frac{\tau}{2}\right) s^*\left(t - \frac{\tau}{2}\right) \exp\{-j\vartheta t\} dt \tag{6.1}$$

which was first derived by Ville and Moyal. Its relation to matched filters was developed by Woodward [190]. Formula (6.1) is traditionally named the *auto-AF*. Accordingly, the *cross-AF* is defined as

$$\text{AF}_{s,g}(\vartheta, \tau) = \int s\left(t + \frac{\tau}{2}\right) g^*\left(t - \frac{\tau}{2}\right) \exp\{-j\vartheta t\} dt \tag{6.2}$$

Unlike the auto-WVD, which is real for any signals, the AF is generally complex valued, i.e.,

$$\text{AF}_{s,g}(\vartheta, \tau) \neq \text{AF}^*_{g,s}(\vartheta, \tau) \tag{6.3}$$

The ambiguity function has been widely used in the context of radar and sonar, and its properties are very well understood. The discussion in this book will be focused only on those aspects that are of importance for a better understanding of Cohen's class.

Based upon the Fourier theorem, for a given ambiguity function $\text{AF}_s(\vartheta, \tau)$, we can compute the time-dependent auto-correlation function via the inverse Fourier transform, i.e.,

$$\frac{1}{2\pi} \int \text{AF}_s(\vartheta, \tau) \exp\{j\vartheta t\} d\vartheta = s\left(t + \frac{\tau}{2}\right) s^*\left(t - \frac{\tau}{2}\right) \tag{6.4}$$

Substituting (6.4) into (5.9) yields

$$\text{WVD}_s(t, \omega) = \frac{1}{2\pi} \iint \text{AF}_s(\vartheta, \tau) \exp\{-j[\omega\tau - \vartheta t]\} d\vartheta d\tau \tag{6.5}$$

which indicates that the Wigner-Ville distribution is a double Fourier transformation[*] of the symmetric ambiguity function. Fig. 6–1 illustrates the relations between the Wigner-Ville distribution and the symmetric ambiguity function.

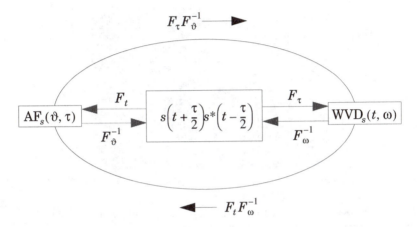

Fig. 6–1 Relationship between Wigner-Ville distribution and ambiguity functions.

Example 6–1   AF of a Gaussian function

$$s(t) = \left(\frac{\alpha}{\pi}\right)^{\frac{1}{4}} \exp\left\{-\frac{\alpha}{2}(t - t_0)^2 + j\omega_0 t\right\} \tag{6.6}$$

The time and frequency centers of $s(t)$ are t_0 and ω_0, respectively. The corresponding symmetric ambiguity function is

$$\text{AF}_s(\vartheta, \tau) = \exp\left\{-\left(\frac{1}{4\alpha}\vartheta^2 + \frac{\alpha}{4}\tau^2\right)\right\} \exp\{j(\omega_0\tau + \vartheta t_0)\} \tag{6.7}$$

Fig. 6–2 illustrates the real part of (6.7), which is centered at the origin (0,0) and oscillates. The phase $\omega_0 t + \vartheta t_0$ is related to the signal's time shift t_0 and frequency modulation ω_0.

---

[*] Strictly speaking, (6.5) contains one Fourier transform and one inverse Fourier transform. For the sake of simplicity, in most literature the operation described by (6.5) is simply remembered as double Fourier transformation.

In contrast, the WVD of the Gaussian function in (6.6) is

$$\text{WVD}_s(t, \omega) = 2\exp\left\{-\frac{1}{\alpha}(t - t_0)^2 - \alpha(\omega - \omega_0)^2\right\} \tag{6.8}$$

which are centered at (t_0, ω_0). In other words, the signal's time shift and frequency modulation are associated with the geological center of the WVD.

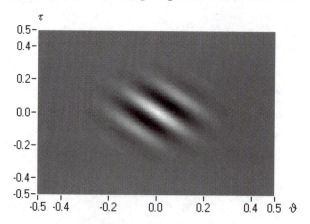

Fig. 6–2 Ambiguity function of the Gaussian signal (real parts).

Example 6–2   AF of two Gaussian functions

$$s(t) = \sum_{i=1}^{2} s_i(t) = \sum_{i=1}^{2}\left(\frac{\alpha}{\pi}\right)^{\frac{1}{4}}\exp\left\{-\frac{\alpha}{2}(t - t_i)^2 + j\omega_i t\right\} \tag{6.9}$$

The two Gaussian functions are concentrated in (t_1, ω_1) and (t_2, ω_2), respectively. The corresponding symmetric ambiguity function is

$$\text{AF}_s(\vartheta, \tau) = \sum_{i=1}^{2}\text{AF}_{s_i}(\vartheta, \tau) + \text{AF}_{s1, s2}(\vartheta, \tau) + \text{AF}_{s2, s1}(\vartheta, \tau) \tag{6.10}$$

where $\text{AF}_{si}(\vartheta, \tau)$ are described by (6.7), which are all concentrated in the origin $(0,0)$. $\text{AF}_{s1, s2}(\vartheta, \tau)$, is

$$\text{AF}_{s1, s2}(\vartheta, \tau) = \exp\left\{-\frac{1}{4\alpha}(\vartheta - \omega_d)^2 + \frac{\alpha}{4}(\tau - t_d)^2\right\} \tag{6.11}$$

$$\times \exp\{j(\omega_\mu \tau - \vartheta t_\mu + \omega_d t_\mu)\}$$

where

$$t_\mu = \frac{t_1 + t_2}{2} \qquad \omega_\mu = \frac{\omega_1 + \omega_2}{2} \qquad t_d = t_1 - t_2 \qquad \omega_d = \omega_1 - \omega_2 \qquad (6.12)$$

Eq.(6.11) indicates that $\mathrm{AF}_{s1,s2}(\vartheta,\tau)$ is concentrated in $(t_1{-}t_2, \omega_1{-}\omega_2)$, away from the origin. $\mathrm{AF}_{s2,s1}(\vartheta,\tau)$ has a similar form to $\mathrm{AF}_{s1,s2}(\vartheta,\tau)$ except for the center in $(t_2{-}t_1, \omega_2{-}\omega_1)$. Fig. 6–3 sketches the locations of each individual term in (6.10).

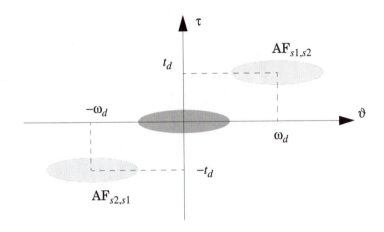

Fig. 6–3 Ambiguity function of two Gaussian signals.

Recall that the cross-WVD has a form

$$\mathrm{WVD}_{s1,s2}(t, \omega) = 2\exp\left\{-\alpha(t - t_\mu)^2 - \frac{1}{\alpha}(\omega - \omega_\mu)^2\right\} \qquad (6.13)$$

$$\times \exp\{j[(\omega - \omega_\mu)t_d + \omega_d t]\}$$

If we write both the Wigner-Ville distribution and symmetric ambiguity function in terms of magnitude and phase, i.e.,

$$\mathrm{WVD}_{s1,s2}(t, \omega) = A_{\mathrm{WVD}}(t, \omega)\exp\{j\varphi_{\mathrm{WVD}}(t, \omega)\} \qquad (6.14)$$

and

$$\mathrm{AF}_{s1,s2}(\vartheta, \tau) = A_{\mathrm{AF}}(\vartheta, \tau)\exp\{j\varphi_{\mathrm{AF}}(\vartheta, \tau)\} \qquad (6.15)$$

then

$$\frac{\partial}{\partial\vartheta}\varphi_{\mathrm{AF}}(\vartheta, \tau) = -t_\mu \qquad \frac{\partial}{\partial\tau}\varphi_{\mathrm{AF}}(\vartheta, \tau) = \omega_\mu \qquad (6.16)$$

which says that the partial derivatives of the phase of the symmetric ambiguity function are equal to the time and frequency center of the Wigner-Ville distribution. Conversely,

$$\frac{\partial}{\partial \omega}\varphi_{\text{WVD}}(t, \omega) = t_d \qquad \frac{\partial}{\partial t}\varphi_{\text{WVD}}(t, \omega) = \omega_d \tag{6.17}$$

which says that the partial derivatives of the phase of the Wigner-Ville distribution are equal to the center of the symmetric ambiguity function. Because the derivative of the phase is usually considered as the frequency, the location of the symmetric ambiguity function directly relates to the rate of oscillation of the Wigner-Ville distribution. In general, the cross-term in the Wigner-Ville domain is severely oscillated, which implies that the partial derivative of the phase of the Wigner-Ville distribution is larger. In the ambiguity domain, this is equivalent to the corresponding ambiguity function being away from the origin. The farther away $\text{AF}_{s1,s2}(\vartheta,\tau)$ is from the origin, the higher the oscillation of $\text{WVD}_{s1,s2}(t,\omega)$.

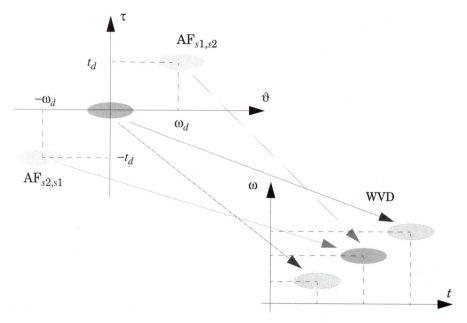

Fig. 6–4 Mapping between AF and WVD.

Because the highly oscillated $\text{WVD}_{s1,s2}(t,\omega)$ has a very small average (thereby having negligible contributions to the useful properties), equivalently, the $\text{AF}_{s1,s2}(\vartheta,\tau)$ that is away from the origin can often be ignored. Fig. 6–4 demonstrates mapping between the ambiguity and Wigner-Ville domains.

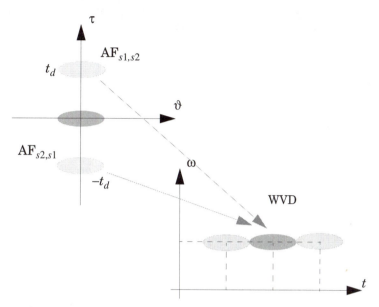

Fig. 6–5 The cross-AFs centered in the τ-axis correspond to the cross-term in the WVD, which oscillates in frequency.

If $\omega_1 = \omega_2 = \omega_0$, that is, the two Gaussian functions have the same frequency center, then $\text{AF}_{s1,s2}(\vartheta,\tau)$ in (6.11) reduces to

$$\text{AF}_{s1,s2}(\vartheta, \tau) = \exp\left\{-\left[\frac{1}{4\alpha}\vartheta^2 + \frac{\alpha}{4}(\tau - t_d)^2\right]\right\}\exp\{j(\omega_0\tau - \vartheta t_\mu)\} \qquad (6.18)$$

which is concentrated in the τ-axis. The distance between the center of $\text{AF}_{s1,s2}(\vartheta,\tau)$ to origin is $|t_d|$, as shown in Fig. 6–5. If $|t_d| \gg 0$, the $\text{AF}_{s1,s2}(\vartheta,\tau)$ corresponds to the $\text{WVD}_{s1,s2}(t,\omega)$ that strongly oscillates in the frequency domain and thereby has a negligible average.

Similarly, if $t_1 = t_2 = t_0$, that is, the two Gaussian functions have the same time center, then $\text{AF}_{s1,s2}(\vartheta,\tau)$ in (6.11) reduces to

$$\text{AF}_{s1,s2}(\vartheta, \tau) = \exp\left\{-\left[\frac{1}{4\alpha}(\vartheta - \omega_d)^2 + \frac{\alpha}{4}\tau^2\right]\right\}\exp\{j[\omega_\mu\tau - (\vartheta - \omega_d)t_0]\} \qquad (6.19)$$

which is concentrated in the ϑ-axis, as illustrated in Fig. 6–6. The distance between the center of $\text{AF}_{s1,s2}(\vartheta,\tau)$ to origin is $|\omega_d|$. If $|\omega_d| \gg 0$, the $\text{AF}_{s1,s2}(\vartheta,\tau)$ corresponds to the $\text{WVD}_{s1,s2}(t,\omega)$ that strongly oscillates in time with the frequency ω_d. Because of the relatively small average, such $\text{WVD}_{s1,s2}(t,\omega)$ has negligible contributions to useful properties.

In order to maintain the useful properties of the WVD, many joint time-frequency representations tend to retain all portions of $\text{AF}(\vartheta,\tau)$ that are in the τ-axis or ϑ-axis. As studied earlier, however, the portions of AF(ϑ,τ) in the τ-axis or ϑ-axis could cause significant undesired terms in the Wigner-Ville domain. On the other hand, when AF($\vartheta,0$) and AF($0,\tau$) are far away from the origin, their contribution to the useful properties is limited. We shall discuss this problem in more detail in subsequent sections.

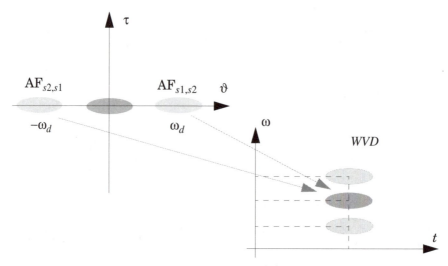

Fig. 6–6 The cross-AFs centered in the ϑ-axis correspond to the cross-term in the WVD, which oscillates in time.

Finally, like the Wigner-Ville distribution, the ambiguity function is also closely related to the wavelet transform, e.g.,

$$\text{CWT}(a,b)\text{CWT}^*(a,-b) = \frac{1}{a}\iint \text{AF}_s(\tau,v)\text{AF}_\psi\left(\frac{\tau-2b}{a}, av\right)d\tau dv \qquad (6.20)$$

PROOF

The right side of (6.20) is equal to

$$\frac{1}{a}\iiint s\left(x+\frac{\tau}{2}\right)s^*\left(x-\frac{\tau}{2}\right)e^{-jvx}dx\int\psi^*\left(y+\frac{\tau-2b}{2a}\right)\psi\left(y-\frac{\tau-2b}{2a}\right)e^{javy}dydtdv$$

$$= \frac{1}{a}\iint s\left(ay+\frac{\tau}{2}\right)s^*\left(ay-\frac{\tau}{2}\right)\psi^*\left(y+\frac{\tau-2b}{2a}\right)\psi\left(y-\frac{\tau-2b}{2a}\right)dyd\tau \qquad (6.21)$$

Substitute

$$y' = ay + \frac{\tau}{2} \qquad \text{and} \qquad \tau' = ay - \frac{\tau}{2} \qquad (6.22)$$

into (6.21). Because

$$dy d\tau = |J| dy' d\tau' = dy' d\tau' \tag{6.23}$$

where $|J|$ denotes the Jacobian determinant (see more detail about multi-variable calculus in Section 5.4), (6.21) becomes

$$\frac{1}{a} \iint s(y') s^*(\tau') \psi^* \left(\frac{y'-b}{a} \right) \psi \left(\frac{y'+b}{a} \right) dy' d\tau' = \text{CWT}(a, b) \text{CWT}^*(a, -b)$$

6.2 Cohen's Class

So far, we have studied the STFT spectrogram, Wigner-Ville distribution, and symmetric ambiguity function. In addition to these three bilinear joint time-frequency representations, we can list at least another dozen counterparts. In 1966, Cohen employed characteristic functions and operator theory to derive a general class of joint time-frequency representations [26]. It can be shown that all these bilinear representations can be written in one general form that is traditionally named *Cohen's class*.

According to [26], the time-dependent auto-correlation function is defined as

$$R(t, \tau) = \frac{1}{2\pi} \int \text{AF}(\vartheta, \tau) \Phi(\vartheta, \tau) \exp\{j\vartheta t\} d\vartheta \tag{6.24}$$

where $\text{AF}(\vartheta, \tau)$ is the symmetric ambiguity function, defined by (6.1), and $\Phi(\vartheta, \tau)$ is called the *kernel function*. Based on the convolution theorem, we can rewrite (6.24) as

$$R(t, \tau) = F^{-1}[\text{AF}(\vartheta, \tau)] \otimes F^{-1}[\Phi(\vartheta, \tau)]$$

$$= \left[s\left(t + \frac{\tau}{2} \right) s^*\left(t - \frac{\tau}{2} \right) \right] \otimes \phi(t, \tau)$$

$$= \int s\left(u + \frac{\tau}{2} \right) s^*\left(u - \frac{\tau}{2} \right) \phi(t - u, \tau) du \tag{6.25}$$

where $\phi(t, \tau)$ denotes the inverse Fourier transformation of $\Phi(\vartheta, \tau)$. Eq.(6.25) says that the general time-dependent auto-correlation function proposed by Cohen is a time-domain linear-filtered function $s(t+\tau/2)s^*(t-\tau/2)$ that is the auto-correlation function employed in the Wigner-Ville distribution. Therefore, the difference between the Wigner-Ville distribution and the member of Cohen's class, such as Choi-Williams distribution and cone-shaped distribution, is completely determined by the nature of the filter $\phi(t,\tau)$. When $\phi(t,\tau)$ is allpass, that is, $\Phi(\vartheta,\tau) = 1$, $R(t,\tau)$ becomes

$$R(t, \tau) = \frac{1}{2\pi}\int AF(\vartheta, \tau)\exp\{j\vartheta t\}d\vartheta = s\left(t + \frac{\tau}{2}\right)s^*\left(t - \frac{\tau}{2}\right) \tag{6.26}$$

which is exactly the Wigner-Ville auto-correlation function. When the kernel function $\Phi(\vartheta,\tau)$ is a valid ambiguity function of an arbitrary time function $\gamma(t)$, then $C(t,\omega)$ becomes the STFT spectrogram with the window function $\gamma(t)$. We shall show this shortly.

Table 6–1

	Properties	Kernel
1	time-shift invariant	independent of time variable t
2	frequency-shift invariant	independent of frequency variable ω
3	realness	$\Phi(\vartheta,\tau) = \Phi(-\vartheta,-\tau)$
4	time marginal	$\Phi(\vartheta,0) = 1$
5	frequency marginal	$\Phi(0,\tau) = 1$
6	instantaneous frequency property	$\Phi(\vartheta,0) = 1$ and $\left.\frac{\partial}{\partial\tau}\Phi(\vartheta, \tau)\right|_{\tau = 0} = 0$
7	group delay property	$\Phi(0,\tau) = 1$ and $\left.\frac{\partial}{\partial\vartheta}\Phi(\vartheta, \tau)\right|_{\vartheta = 0} = 0$
8	positivity	$\Phi(\vartheta,\tau)$ is an ambiguity function of $\gamma(t)$ or signal-dependent

Substituting (6.24) into (5.3) obtains the form of Cohen's class

$$C(t, \omega) = \frac{1}{2\pi}\iint AF(\vartheta, \tau)\Phi(\vartheta, \tau)\exp\{j(\vartheta t - \omega\tau)\}d\vartheta d\tau \tag{6.27}$$

Alternatively, replacing $R(t,\tau)$ in (5.3) by (6.25) yields

$$C(t, \omega) = \iint s\left(u + \frac{\tau}{2}\right)s^*\left(u - \frac{\tau}{2}\right)\phi(t - u, \tau)du\exp\{-j\omega\tau\}d\tau \tag{6.28}$$

To compute the ambiguity function we have to have the entire time record, which often is impossible in practice. Because the formula (6.28) does not need to compute the ambiguity functions, it is suitable for real implementations.

The significance of Cohen's work is to reduce the problem of the design of the time-dependent spectrum to the selection of the kernel function $\Phi(\vartheta,\tau)$. We list the relations between the kernel functions and useful properties in Table 6–1 without proofs. Readers interested in derivations are encouraged to consult [32].

In general, Cohen's class could be negative unless the kernel function is signal-dependent or it is the ambiguity function of a function $\gamma(t)$. In this case, $C(t,\omega)$ is equivalent to the STFT spectrogram, the square of the STFT.

PROOF

Let

$$\Phi(\vartheta, \tau) = \int \gamma\left(t + \frac{\tau}{2}\right)\gamma^*\left(t - \frac{\tau}{2}\right)\exp\{-j\vartheta t\}dt \tag{6.29}$$

Then, its inverse Fourier transform is

$$\phi(t, \tau) = \gamma\left(t + \frac{\tau}{2}\right)\gamma^*\left(t - \frac{\tau}{2}\right) \tag{6.30}$$

Substituting (6.30) into (6.25) obtains

$$C(t, \omega) = \iint \gamma\left(t - u + \frac{\tau}{2}\right)\gamma^*\left(t - u - \frac{\tau}{2}\right)s\left(u + \frac{\tau}{2}\right)s^*\left(u - \frac{\tau}{2}\right)\exp\{-j\omega\tau\}dud\tau$$

$$= \iint s\left(u + \frac{\tau}{2}\right)\gamma^*\left[t - \left(u + \frac{\tau}{2}\right)\right]\exp\left\{-j\omega\left(u + \frac{\tau}{2}\right)\right\}$$

$$s^*\left(u - \frac{\tau}{2}\right)\gamma\left[t - \left(u - \frac{\tau}{2}\right)\right]\exp\left\{j\omega\left(u - \frac{\tau}{2}\right)\right\}dud\tau$$

Replacing (see more detail about multivariable calculus in Section 5.4)

$$x = u + \frac{\tau}{2} \qquad y = u - \frac{\tau}{2} \tag{6.31}$$

yields

$$C(t, \omega) = \iint s(x)\gamma^*[t - x]\exp\{-j\omega y\}s^*(y)\gamma[t - y]\exp\{j\omega y\}dxdy = |S(\omega)|^2$$

which manifestly is non-negative.

Note that when the kernel $\Phi(\vartheta,\tau)$ is the ambiguity function of a function $\gamma(t)$, the resulting $C(t,\omega)$ does not satisfy the properties 4, 5, 6, and 7.

From the classical energy concept, the signal's energy distribution (or energy density function) obviously should be non-negative. Wigner showed (see

[187] and [188]), however, that the bilinear transform cannot satisfy marginal conditions and be non-negative simultaneously.

A natural question at this point is, does a non-negative time-frequency distribution (or time-frequency density function) exist? The answer is *yes*. In fact, if we do not limit ourselves to bilinear transformations, such as Cohen's formulae, (6.27) and (6.25), then we could easily make non-negative time-frequency functions with marginal condition properties. For example, let

$$P(t, \omega) = \frac{|s(t)|^2}{\|s(t)\|^2} |S(\omega)|^2 \tag{6.32}$$

where

$$\|s(t)\|^2 = \int |s(t)|^2 dt = \frac{1}{2\pi} \int |S(\omega)|^2 d\omega$$

Evidently, $P(t,\omega)$ in (6.32) is non-negative and satisfies time/frequency marginal conditions. With a normalization, it could be formulated as a joint density function. But such a time-frequency density function is *meaningless* for time-frequency analysis, because it does not convey any information regarding a signal's local behaviors.

Then, the next question will be, does a *meaningful* non-negative time-frequency distribution exist? Unfortunately, the answer of this question is, "*we don't know.*" Although there are many ways of creating non-negative time-frequency functions, none of them has been proved to truly reflect a signal's time-varying nature.

The subject of the existence of a non-negative time-frequency distribution that also reflects the *signal's time-varying nature* so far remains a research topic. It not only is of interest to signal processing, but also is fundamental in physics. The reader interested in this topic is encouraged to consult the papers by Cohen and Posch [27] and Janssen (see [87], [89], and [92]).

All discussions in this section are primarily focused on the continuous-time cases and therefore all conclusions only hold for the continuous-time signals. The discrete counterparts were studied by Jeong and Williams [96], Morris and Wu [128], and Kootsookos et al. [106].

6.3   Some Members of Cohen's Class

One of the main motivations for studying Cohen's class for the past ten years has been to seek a time-dependent spectrum that not only preserves all useful properties, but also has reduced cross-term interference. As discussed in the previous section, the portion of the ambiguity function that corresponds to auto-terms is always connected to the origin, whereas the part of the ambiguity function that is related to the cross-terms tends to spread somewhere else. This observation inspires many studies to search for a kernel function $\Phi(\vartheta,\tau)$ such that the product $|\Phi(\vartheta,\tau)AF(\vartheta,\tau)|$ is enhanced in the vicinity of origin and suppressed everywhere else. At the same time, $\Phi(\vartheta,\tau)$ should satisfy as many properties listed in Table 6–1 as possible.

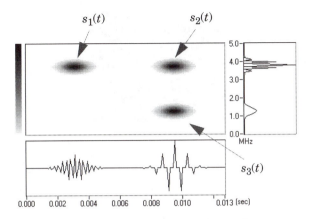

Fig. 6–7 Three-tone test signal.

Fig. 6–7 illustrates a three-tone test signal. The signals $s_1(t)$ and $s_2(t)$ have the same frequency center. The signals $s_3(t)$ and $s_2(t)$ have the same time center. Fig. 6–8 plots the WVD of the test signal, which indicates that each pair of auto-terms creates one cross-term.

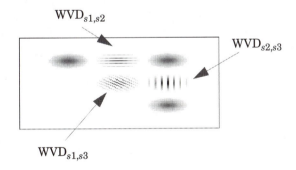

Fig. 6–8 WVD of the three-tone test signal.

Fig. 6–9 depicts the three-tone test signal in the ambiguity domain. It shows that, except for the cluster centered in (0,0), all other signals are caused by cross-terms.

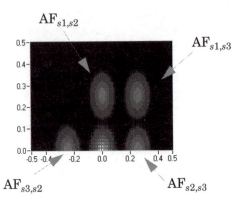

Fig. 6–9 Ambiguity function of the three-tone test signal.

As discussed previously, the useful properties and cross-term suppression cannot be accomplished simultaneously. For example, to reduce the cross-term interference, the product $|\phi(\vartheta,\tau)AF(\vartheta,\tau)|$ has to vanish for larger ϑ and τ. On the other hand, to preserve the time and frequency marginal conditions, the following must hold:

$$\phi(\vartheta, 0)AF(\vartheta, 0) = AF(\vartheta, 0) \qquad \phi(0, \tau)AF(0, \tau) = AF(0, \tau) \qquad (6.33)$$

which implies that all portions of $AF(\vartheta,\tau)$ in both the τ-axis and ϑ-axis have to be kept, no matter how far they are from the origin. The direct consequence is that the resulting representation will preserve all cross-terms that are created by two functions that have either the same time center or frequency center.

In what follows, we shall investigate the performances of the three most popular members of Cohen's class by the test signal introduced above.

6.3.1 Choi-Williams Distribution

To suppress the portions of the AF that are away from the origin, Choi and Williams [23] introduced the exponential kernel as

$$\Phi(\vartheta, \tau) = \exp\{-\alpha(\vartheta\tau)^2\} \qquad (6.34)$$

It is rather easy to check that the exponential kernel in (6.34) satisfies all properties listed in Table 6–1 except for the positivity. Moreover, $\Phi(0,0) = 1$ and $\Phi(\vartheta,\tau) < 1$ for $\vartheta \neq 0$ and $\tau \neq 0$, which imply that the exponential kernel will suppress the cross-terms created by two functions that have both different time and frequency centers. The parameter α controls the decay speed, as shown in Fig. 6–10. The larger the α is, the more the cross-terms are suppressed. On the other hand, the larger the α, the more the auto-terms are affected. Therefore, there is a trade-off for the selection of the parameter α.

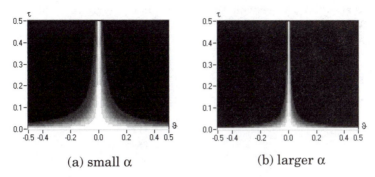

(a) small α (b) larger α

Fig. 6–10 Exponential kernel with small α (left) and larger α (right).

The inverse Fourier transform of the exponential kernel in (6.34) is

$$\phi(t, \tau) = \frac{1}{\sqrt{4\pi\alpha\tau^2}} \exp\left\{-\frac{1}{4\alpha\tau^2}t^2\right\} \tag{6.35}$$

Substituting (6.35) into (6.25) yields

$$\mathrm{CWD}(t, \omega) = \iint \frac{1}{\sqrt{4\pi\alpha\tau^2}} \exp\left\{-\frac{(t-u)^2}{4\alpha\tau^2}\right\} s\left(u + \frac{\tau}{2}\right) s^*\left(u - \frac{\tau}{2}\right) \exp\{-j\omega\tau\} du\, d\tau \tag{6.36}$$

which is commonly named the *Choi-Williams distribution* (CWD) .

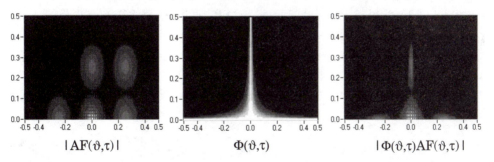

$|\mathrm{AF}(\vartheta,\tau)|$ $\Phi(\vartheta,\tau)$ $|\Phi(\vartheta,\tau)\mathrm{AF}(\vartheta,\tau)|$

Fig. 6–11 Cross-term suppression by exponential kernel. (Although the exponential kernel suppresses the portion that is away from the origin, it preserves all cross-terms that are in the ϑ-axis and τ-axis.)

Fig. 6–11 depicts the process of cross-term suppression by exponential kernel function for the three-tone test signal. Although the exponential kernel suppresses the portion that is away from the origin, it preserves all cross-terms that are in the ϑ-axis and τ-axis. Consequently, the CWD contains strong horizontal and vertical ripples. The horizontal ripples are caused by the auto-terms that have the same frequency center; the vertical ripples correspond to the auto-terms that have the same time center. Fig. 6–12 illustrates the CWD of the three-tone test signal. Moreover, the reader should bear in mind that theoretically, the CWD preserves the properties of the Wigner-Ville distribution, but it is not the case for the discrete-time signal.

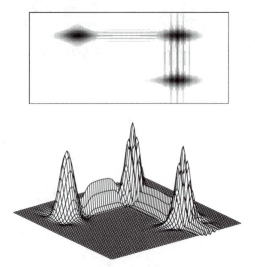

Fig. 6–12 CWD of the three-tone test signal.

6.3.2 Cone-Shape Distribution

Another popular representation is the cone-shape distribution that was introduced by Zhao, et al [197]. The cone-shape function can be expressed as

$$\phi(t, \tau) = \begin{cases} g(\tau) & |\tau| \geq 2|t| \\ 0 & \text{otherwise} \end{cases} \tag{6.37}$$

which is plotted in Fig. 6–13. In the ambiguity domain, the cone-shape function has the form

$$\Phi(\vartheta, \tau) = g(\tau) \int_{-\tau/2}^{\tau/2} \exp\{-j\vartheta t\}dt = 2g(\tau)\frac{\sin(\vartheta\tau/2)}{\vartheta} \tag{6.38}$$

Let

$$g(\tau) = \frac{1}{\tau}\exp\{-\alpha\tau^2\} \tag{6.39}$$

Then, (6.38) becomes

$$\Phi(\vartheta, \tau) = \frac{\sin(\vartheta\tau/2)}{\vartheta\tau/2}\exp\{-\alpha\tau^2\} \qquad \alpha > 0 \tag{6.40}$$

The parameter α controls the degree of suppression, as shown in Fig. 6–14. The larger the parameter α, the more the cross-terms are suppressed (at the expense of more disturbed auto-terms). Obviously,

$$\Phi(\vartheta, \tau) = \begin{cases} 1 & \tau = 0 \\ \exp\{-\alpha\tau^2\} & \vartheta = 0 \end{cases} \tag{6.41}$$

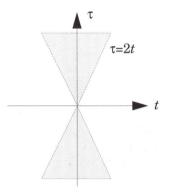

Fig. 6–13 Cone-shape function

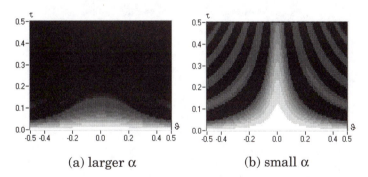

(a) larger α (b) small α

Fig. 6–14 Cone-shape kernel in ambiguity domain.

Unlike the exponential kernel function (6.35), which preserves all portions of AF in both the ϑ-axis as well as τ-axis, the cone-shape kernel (6.40) suppresses those AF which are in the τ-axis. Consequently, the cone-shape distribution is able to effectively attenuate the cross-terms created by two functions that have the same frequency center. Fig. 6–15 plots $|\text{AF}(\vartheta,\tau)|$ of the three-tone test signal, $\Phi(\vartheta,\tau)$ of the cone-shape function, and $|\Phi(\vartheta,\tau)\text{AF}(\vartheta,\tau)|$. Although the cross-terms in the ϑ-axis are preserved, the cross-term in the τ-axis is suppressed.

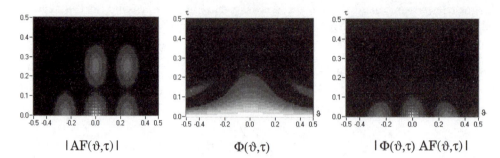

$$|AF(\vartheta,\tau)| \qquad\qquad \Phi(\vartheta,\tau) \qquad\qquad |\Phi(\vartheta,\tau)\,AF(\vartheta,\tau)|$$

Fig. 6–15 Cross-term suppression by cone-shape kernel. (Although the cross-terms in the ϑ-axis are preserved, the cross-term in the τ-axis is suppressed.)

Fig. 6–16 depicts the cone-shape distribution for the three-tone test signal. Compared to the Wigner-Ville distribution in Fig. 6–8 and the Choi-Williams distribution in Fig. 6–12, the cross-term interference is significantly reduced in the cone-shape distribution. In particular, the cross-term between two that have the same frequency center is completely removed. However, the interference created by two functions with the same time center still exists.

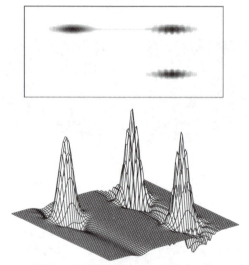

Fig. 6–16 Cone-shape distribution.

In the area of speech analysis, the cone-shape distribution has been found to be a favorite alternative for the conventional STFT spectrogram. Fig. 6–17 illustrates analysis results of an utterance by a five-year-old boy. The top plot is computed by the cone-shape distribution, which not only clearly demonstrates the formant patterns, but also depicts the vertical structures well. Fig. 6–18 plots the same speech signal computed by wideband and narrowband STFT spectrogram, respectively. While the format in the wideband STFT spectrogram widely spreads out, it is mixed with harmonics in the narrowband STFT spectrogram.

Cone-shape Distribution

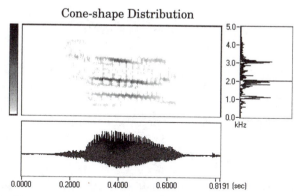

Fig. 6–17 "Wood" spoken by a five-year-old boy. (Data courtesy of Y. Zhao, the Beckman Institute at the University of Illinois.)

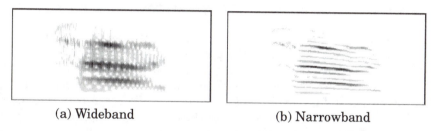

(a) Wideband   (b) Narrowband

Fig. 6–18 Wideband and narrowband STFT spectrogram. (While the format in the wideband STFT spectrogram widely spreads out, it is mixed with harmonics in the narrowband STFT spectrogram.)

6.3.3 Signal-Dependent Time-Frequency Representations[*]

Unlike the previous kernels, which emphasize preserving the properties of the Wigner-Ville distribution over matching the shape of auto-components, the signal-dependent kernels aim to optimally pass the auto-components while suppressing cross-components. Since a fixed kernel acts on the ambiguity function as a mask or filter, it is limited in its ability to perform this function. The locations of the auto- and cross-components depend on the signal to be analyzed; hence we expect to obtain good performance for a broad class of signals only by using a *signal-dependent* kernel. To find the bilinear time-frequency distribution that provides in some sense the "best" time-frequency representations (TFD) for a given signal, Baraniuk and Jones have formulated the signal-dependent kernel design procedure as an optimization problem (see [6] and [7]).

[*] Contributed by Richard G. Baraniuk, Department of Electrical and Computer Engineering, Rice University, Houston, Texas 77251-1892.

The 1/0 kernel method [7]

Given a signal and its AF, the optimal 1/0 kernel is defined as the real, non-negative function ϕ_{opt} that solves the following optimization problem

$$\max_{\phi} \int\int |AF(\vartheta, \tau)\Phi(\vartheta, \tau)|^2 d\vartheta d\tau \tag{6.42}$$

subject to

$$\Phi(0, 0) = 1 \tag{6.43}$$

$$\Phi(\vartheta, \tau) \text{ is radially non-increasing} \tag{6.44}$$

$$\int\int |\Phi(\vartheta, \tau)|^2 dv d\vartheta \leq 0 \tag{6.45}$$

the radially non-increasing constraint (6.45) can be expressed explicitly as

$$\Phi(\gamma_1, \psi) \geq \Phi(\gamma_2, \psi) \qquad \forall \ \gamma_1 \leq \gamma_2 \qquad \forall \ \psi$$

where γ_i and ψ correspond to the polar coordinates radius and angle, respectively.

The constraints (6.43) to (6.45) and performance measure (6.42) are formulated so that the optimal kernel (OK) passes auto-components and suppresses cross-components. The constraints force the optimal kernel to be a lowpass filter of fixed volume α; maximizing the performance measure encourages the passband of the kernel to lie over the auto-components. Both the performance measure and the constraints are insensitive to the orientation angle and aspect ratio (scaling) of the signal components in the (ϑ, τ) plane.

By controlling the volume under the optimal kernel, the parameter α controls the trade-off between cross-component suppression and smearing of the auto-components. Reasonable bounds are $1 \leq \alpha \leq 5$. At the lower bound, the optimal kernel shares the same volume as an STFT spectrogram kernel, while at the upper bound, the optimal kernel smooths only slightly. In fact, as α, the optimal kernel distribution converges to the WVD of the signal.

Analyzing a signal with an optimal-kernel distribution requires a three-step procedure: (1) compute the AF of the signal; (2) solve the linear program (6.42) to (6.45) in variables $|\phi|^2$ (a fast algorithm is given in [5]); (3) Fourier transform the AF-kernel product $AF(\vartheta, \tau)\Phi(\vartheta, \tau)$.

The radially Gaussian kernel method [6]

Although the 1/0 kernel is optimal according to the criteria (6.42) to (6.45), its sharp cutoff may introduce ringing into the OK distribution, especially for small values of the kernel volume parameter α. For an alternative, direct approach to smooth optimal kernels, explicit smoothness constraints can be appended to the kernel optimization formulation (6.42) to (6.45). In [6], the kernel is constrained to be Gaussian along radial profiles

$$\Phi(\vartheta, \tau) = \exp\left\{-\frac{\vartheta^2 + \tau^2}{2\sigma^2(\psi)}\right\} \qquad (6.46)$$

The term $\sigma(\psi)$ represents the dependence of the Gaussian spread on radial angle $\psi = \arctan(\tau/\vartheta)$. Any kernel of the form (6.46) is bounded and radially non-increasing and, furthermore, smooth if σ is smooth. Since the shape of a radially Gaussian kernel is completely parameterized by this function, finding the optimal, radially Gaussian kernel for a signal is equivalent to finding the optimal function σ_{opt} for the signal. A gradient ascent/Newton algorithm solving the (non-linear) system (described by (6.42), (6.45), and (6.46)) is detailed in [6].

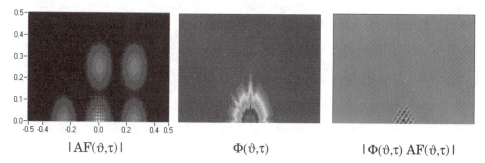

$$|AF(\vartheta, \tau)| \qquad\qquad \Phi(\vartheta, \tau) \qquad\qquad |\Phi(\vartheta, \tau)\, AF(\vartheta, \tau)|$$

Fig. 6–19 Optimal kernel and $|AF(\vartheta,\tau)\Phi(\vartheta,\tau)|$ of the three-tone test signal.

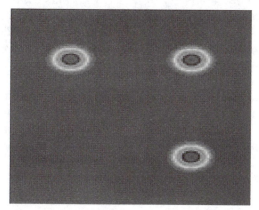

Fig. 6–20 Optimal-kernel distribution for the three-tone test signal.

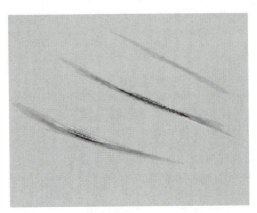

Fig. 6–21 Bat example. (Bat data provided by Curtis Condon, Ken White, and Al Feng of the Beckman Institute at the University of Illinois.)

Fig. 6–19 shows the AF domain optimal kernel and product $|AF(\vartheta,\tau)\Phi(\vartheta,\tau)|$ for the three-tone test signal. Fig. 6–20 shows the optimal-kernel distribution. As another example, Fig. 6–21 shows a time-frequency analysis of 2.5 ms of an echo-location pulse emitted by the large brown bat, *Eptesicus fuscus*.

Extensions

Additional constraints The goal of the optimization problems (6.42) to (6.45) and (6.42), (6.45), (6.46) is strictly to find the kernels that optimally pass auto-components and suppress cross-components. However, additional constraints encouraging additional kernel properties can also be considered. An attractive feature of the optimal-kernel design formulation is the ease with which it can be customized to incorporate application-specific knowledge into the design process. For example, constraints have been developed to coerce OK TFDs to satisfy the marginal distributions and/or preserve the time support of the signal (see [6] and [7]).

Adaptive formulations While the 1/0 and radially Gaussian distributions generally perform well, they are block-oriented techniques that design only one kernel for the entire signal. For analyzing signals with characteristics that change over time, for real-time, on-line operation, or for very long signals, adaptive signal-dependent TFDs are required.

Adaptation of the kernel to track the local signal characteristics over time requires that the kernel optimization procedure consider only the local signal characteristics. An ambiguity-domain design procedure such as the radially Gaussian kernel optimization technique described above does not immediately admit such time localization, because the AF includes information from all times and frequencies in the signal. This difficulty has been surmounted, however, by the development of a time-localized, or short-time, AF [99]. Application of the radially Gaussian kernel optimization procedure to the short-time AF localized at time t_0 produces an optimal kernel $\Phi_{opt}(\vartheta, \tau; t_0)$ and an optimal-kernel distribution frequency slice $C_{opt}(t_0, f)$ at t_0. Since the algorithm alters the kernel at each time to achieve optimal local performance, better tracking of the signal changes results. A simpler adaptive algorithm based on the short-time Fourier transform is derived in [98].

Summary

Cohen's class provides a general form of time-dependent auto-correlation function for the bilinear transformation. The significance of Cohen's works were to reduce the design of the desired time-dependent spectrum $C(t, \omega)$ to the selection of the kernel function $\Phi(\vartheta, \tau)$. The Wigner-Ville distribution not only has the simplest kernel $\Phi(\vartheta, \tau) = 1$, but also possesses the most useful properties. One of the major concerns in applications has been the cross-term interference. Because the interference usually is associated with high oscillations, the natural way of reducing it is to make $\Phi(\vartheta, \tau)$ lowpass. However, there is a trade-off; to precisely preserve all useful properties, one has to tolerate a significant amount of interference.

The original Cohen's class was developed for continuous-time functions. All relationships between the kernel function $\Phi(\vartheta, \tau)$ and Cohen's class $C(t, \omega)$ presented in this chapter only apply for continuous-time signals. Those results cannot be duplicated in discrete-time cases by simply replacing the continuous variables t and ω with discrete quantities.

Cohen's class is a typical linear filtering. As an alternative, in Chapter 7, we shall introduce a non-linear filtering approach, time-frequency distribution series, to reduce interference.

CHAPTER 7

Time-Frequency Distribution Series

*T*he main deficiency of the Wigner-Ville distribution is the so-called cross-term interference. At any time instant, if there is more than one frequency tone, then the Wigner-Ville distribution may become messed up because of the presence of undesired terms. However, the cross-terms highly oscillate and are localized, which always occur in the midway of the pair of corresponding auto-terms. On the other hand, useful properties, such as the time marginal, frequency marginal, and instantaneous frequency property, are all obtained by *averaging* the Wigner-Ville distribution. For example, the time marginal condition is computed by integrating (*averaging*) the Wigner-Ville distribution over frequency. This observation suggests that if the Wigner-Ville distribution can be decomposed as the sum of *two-dimensional* (*time and frequency*) *localized harmonic functions*, such as the 2D Gabor expansion, then we could use the low-order harmonic terms to delineate the time-dependent spectrum with reduced interference. High-order harmonics that introduce high oscillation have relatively small averages and thereby have negligible influence on the useful properties. The signal energy and useful properties are mainly determined by a few low-order harmonic terms.

In Section 7.1, we discuss the motivation of the decomposition of the Wigner-Ville distribution. In principle, such decomposition can be realized by the 2D Gabor expansion. From the implementation point of view, however, the 2D Gabor expansion is not an ideal choice. The major difficulty lies in the evaluation of expansion coefficients. Unlike the one-dimensional (1D) Gabor expansion, in which the dual function can be obtained with the help of elementary linear algebra, the computation of the dual function for a given 2D Gabor elementary function in general would be much more involved. In particular, the 2D Gabor coefficient is the 2D inner product between the dual function and the Wigner-Ville distribution, which is computationally expensive and requires knowing the complete Wigner-Ville distribution in advance.

Taking advantage of the fact that the Wigner-Ville distribution of a 1D Gaussian function is a 2D Gaussian function, in Section 7.2, we introduce a 1D Gabor expansion-based decomposition scheme for the Wigner-Ville distribution. The resulting representation is called a *time-frequency distribution series,* $\text{TFDS}_D(t,\omega)$, where D denotes the order of the time-frequency distribution series. Because the time-frequency distribution series is the Gabor expansion- and WVD-based time-dependent spectrum, it is also known as *Gabor spectrogram* in the industry.

In Section 7.3, we discuss the discrete version of the time-frequency distribution series. Because the continuous-time time-frequency distribution series has a closed form for the arbitrary signals, the discrete time-frequency distribution series can be obtained by directly sampling the continuous-time time-frequency distribution series, such as

$$\text{TFDS}_D[i, k] \equiv \text{TFDS}_D(t, \omega)\big|_{t = i\Delta t,\, \omega = 2\pi k / \Delta t}$$

where Δt denotes the sampling interval. Consequently, all properties possessed by the continuous-time time-frequency distribution series are automatically carried over to its discrete counterpart (at least well approximate). There is no aliasing problem such as occurred in the discrete Wigner-Ville distribution.

In Section 7.4, we discuss the problem of the selection of Gabor elementary function $h[k]$ and dual function $\gamma[k]$. In Section 7.5, we address the difference between Cohen's class and the time-frequency distribution series. While Cohen's class represents the linear operation, truncated time-frequency distribution series is the result of a non-linear lowpass filtering.

With the time-frequency distribution series, we gain a better understanding of the nature of the cross-term. The concept of the cross-term in fact is ambiguous. As introduced in Chapter 2, a given signal can be broken up in an infinite number of ways. The different decomposition schemes will lead to the different cross-terms. Hence, cross-terms are not unique. Moreover, the cross-terms are not always *ghosts*. When the cross-term is created by the pair of elementary functions that are close to each other, it has a significant influence on the useful properties and thereby cannot be simply gotten rid of. It is the high harmonics that interferes with the meaningful pattern of time-dependent spectra. The useful properties of the Wigner-Ville distribution are mainly determined by a few lowest harmonics. As shown, the leading terms of the time-frequency distribution series not only well delineate time-dependent power spectra, but also significantly reduce the interference.

The concept of the time-frequency distribution series was first introduced by Qian and Morris [147], in which, however, only the result of the zeroth order time-frequency distribution series was reported. The general case and practical algorithms were further studied by Qian and Chen [152].

7.1 Decomposition of the Wigner-Ville Distribution

The concept of a time-dependent spectrum has proved indispensable in a wide range of applications involving non-stationary signals - from radar and sonar to biomedicine and geophysics. A particularly successful approach to joint time-frequency representation has been driven by the Wigner-Ville distribution. Compared to many other schemes, the Wigner-Ville distribution can better characterize a signal's frequency content changes. For example, the conditional mean frequency of the Wigner-Ville distribution is always equal to the signal's mean instantaneous frequency, the first derivative of the phase. That is, if $s(t) = A(t)e^{j\varphi(t)}$, then

$$\langle \omega \rangle_t = \frac{\int \omega \text{WVD}_s(t, \omega)d\omega}{\int \text{WVD}_s(t, \omega)d\omega} = \frac{1}{2\pi|s(t)|^2}\int \omega \text{WVD}_s(t, \omega)d\omega = \varphi'(t) \qquad (7.1)$$

Note that neither the STFT spectrogram nor scalogram possesses it.

The relation (7.1) has been considered the most important property for a qualified time-dependent spectrum. We often use the conditional mean frequency to evaluate whether or not a proposed time-dependent spectrum reflects a signal's local behaviors.

The main deficiency of the Wigner-Ville distribution is the cross-term interference. As we have shown in the previous chapters, the Wigner-Ville distribution often creates highly oscillated terms in places where they are not expected. It is interesting to observe, however, that the cross-term interference highly oscillates and is localized. It always occurs in the midway of a pair of corresponding auto-terms. On the other hand, the conditional mean frequency (7.1) and other useful properties of the Wigner-Ville distribution, such as time and frequency marginal conditions, depend on the *average* of WVD only. Hence, if the WVD is decomposed as the sum of the *2D localized harmonic functions*, such as the 2D Gaussian functions, then we can use low-order harmonics to delineate the time-dependent spectrum with limited cross-term interference. This is because high harmonic terms have relatively small averages. The useful properties of the WVD are mainly determined by the low-order harmonics. Discarding the high harmonics will remove the high oscillation with negligible effect to the other useful properties. Such a decomposition scheme was first reported in [147] and further elaborated in [152].

The simplest way of decomposing the WVD as the sum of the *2D localized harmonic functions* is to apply a 2D Gabor expansion [114], i.e.,

$$\text{WVD}_s(t, \omega) = \sum_{i, k, p, q} D_{i, k, p, q} H_{i, k, p, q}(t, \omega) \qquad (7.2)$$

where $H(t,\omega)$ is specified as the cross-WVD of two Gaussian functions, i.e.,

$$H_{i,k,p,q}(t,\omega) = \exp\left\{-\alpha(t-iT)^2 - \frac{1}{\alpha}(\omega-k\Omega)^2\right\}\exp\left\{j(pT\omega + q\Omega t - k\Omega pT)\right\} \quad (7.3)$$

When the two Gaussian functions have the same time and frequency centers, $(mT,n\Omega)$, i.e.,

$$h_{i,k}(t) = \exp\left\{-\alpha(t-iT)^2 + jk\Omega\right\} \quad (7.4)$$

then (7.3) reduces to

$$H_{i,k,p,q}(t,\omega) = 2\exp\left\{-\alpha(t-iT)^2 - \frac{1}{\alpha}(\omega-k\Omega)^2\right\} \quad (7.5)$$

which is the joint energy density function of the Gaussian signal of (7.4). The function of (7.5) is real, which is the special case of (7.3). In general, however, the complex exponential part in (7.3) will not vanish.

Because the function defined in (7.3) is an optimally concentrated WVD and related to the signal energy distribution, we consider it an *energy atom*. Fig. 7–1 plots a typical energy atom described by (7.3), which is concentrated, symmetrical, and oscillates in the joint time-frequency domain.

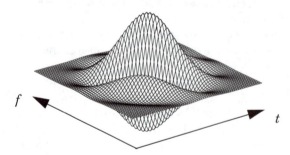

Fig. 7–1 All energy atoms are concentrated, symmetrical, and oscillating in the joint time-frequency domain.

The weight of each individual energy atom is determined by 2D Gabor coefficients $D_{i,k,p,q}$. T and Ω are time and frequency sampling steps. p and q reflect the rate of the oscillation of the energy atoms, in the time and frequency domains, respectively.

Eq. (7.2) demonstrates that the WVD can be constructed in terms of a number of infinite energy atoms $H(t,\omega)$. All these atoms are concentrated, symmetrical, and oscillate in the joint time-frequency domain. In what follows, we show that the contributions of each individual energy atom to the useful properties are inversely proportional to the rate of its oscillation.

In Chapter 5, we proved the time marginal of the WVD is equal to the signal's instantaneous energy, that is,

$$\frac{1}{2\pi}\int \text{WVD}_s(t, \omega)d\omega = |s(t)|^2 \tag{7.6}$$

Substituting (7.2) into (7.6) yields

$$|s(t)|^2 = \sum_p A_p \sum_{i, k, q} D_{i, k, p, q} \exp\{-\alpha(t - iT)^2 + j(q\Omega t + k\Omega pT)\} \tag{7.7}$$

which shows that the time marginal of the WVD consists of the group of 1D time-shifted and frequency-modulated Gaussian functions. The weight of each individual Gaussian function is

$$A_p = \sqrt{\frac{\alpha\pi}{4}}\exp\left\{-\frac{\alpha}{4}(pT)^2\right\} \tag{7.8}$$

which is inversely proportional to $(pT)^2$. In other words, the contribution of each energy atom to the marginal condition exponentially decays as the magnitude of p increases. The higher the harmonics, the less the influence on the time marginal. When all atoms are included, the right side of (7.7) manifestly converges to the signal instantaneous energy $|s(t)|^2$. However, the time marginal of the WVD is mainly determined by a few low harmonic terms.

The reader may verify other properties, such as the conditional mean frequency, by substituting (7.2) into (7.1). It will not be surprising that we will obtain similar observations for all useful properties introduced earlier. We leave this as an exercise for the reader. It is generally true that the high-order harmonics introduce unwanted high oscillation but have limited contributions to the useful properties. The useful properties are mainly determined by lower order harmonic terms. Selectively discarding the high-order harmonics allows us to balance the cross-term interference and useful properties of the WVD.

The remaining problem is the implementation of the 2D Gabor expansion described by (7.2). Based on the expansion theory discussed in Chapter 2, the 2D Gabor coefficients $D_{i,k,p,q}$ could be computed via the regular inner product operation, i.e.,

$$D_{i, k, p, q} = \langle \text{WVD}_s(t, \omega), \gamma_{i, k, p, q}(t, \omega) \rangle \tag{7.9}$$

where $\gamma(t, \omega)$ denotes the dual function of the 2D Gaussian function $H(t, \omega)$ in (7.2).

There are two problems associated with the formula (7.9) from the application point of view. First, unlike the 1D Gabor expansion, it is not trivial, in general, to compute the dual function $\gamma(t, \omega)$ for the given 2D Gabor elementary function and sampling steps, T and Ω. Second, (7.9) is a 2D inner product operation, which is computationally expensive. In particular, it requires knowing the entire WVD in advance. These limitations actually prevent the 2D Gabor expansion from any real applications in which the computational efficiency is a primary issue.

In the next section, we shall introduce a more practical approach of decomposing the WVD. Although the resulting decomposition is not exactly the same as the formula (7.2), the performance has been found very close to the 2D Gabor expansion-based algorithm.

7.2 Time-Frequency Distribution Series

Let's first apply the 1D Gabor expansion to the signal $s(t)$, i.e.,

$$s(t) = \sum_{m=-\infty}^{\infty} \sum_{n=-\infty}^{\infty} C_{m,n} h_{m,n}(t) \tag{7.10}$$

where

$$h \equiv h_{m,n}(t) = \left(\frac{\alpha}{\pi}\right)^{\frac{1}{4}} \exp\left\{-\frac{\alpha}{2}(t-mT)^2 + jn\Omega t\right\} \tag{7.11}$$

The Gabor coefficients $C_{m,n}$ are determined by

$$C_{m,n} = \int s(t)\gamma^*_{m,n}(t)dt = \int s(t)\gamma^*(t-mT)e^{-jn\Omega t}dt = \text{STFT}(mT, n\Omega) \tag{7.12}$$

where $\gamma_{opt}(t)$ denotes the dual function whose shape is optimally close to the Gaussian elementary function $h(t)$. As discussed in Chapter 3, such decomposition is the *orthogonal-like Gabor expansion*. This type of expansion ensures that the Gabor coefficients $C_{m,n}$ indeed reflect the signal behavior in the vicinity of $[mT - \Delta_t, mT + \Delta_t] \times [n\Omega - \Delta_\omega, n\Omega + \Delta_\omega]$. Fig. 7–2 illustrates the Gabor sampling lattice, in which each intersection corresponds to the center of one Gabor elementary function $h_{m,n}(t)$.

Taking the Wigner-Ville distribution with respect to (7.10) yields,

$$\text{WVD}_s(t, \omega) = \sum_{m,n} \sum_{m',n'} C_{m,n} C_{m',n'} \text{WVD}_{h,h'}(t, \omega) \tag{7.13}$$

where

$$\text{WVD}_{h,h'}(t, \omega) = 2\exp\left\{-\alpha\left(t - \frac{m+m'}{2}T\right)^2 - \frac{1}{\alpha}\left(\omega - \frac{n+n'}{2}\Omega\right)^2\right\} \tag{7.14}$$

$$\exp\left\{-j\frac{n+n'}{2}\Omega(m-m')T\right\}\exp\{j[(m-m')T\omega + (n-n')\Omega t]\}$$

If we let

$$\frac{m + m'}{2} = i \qquad \frac{n + n'}{2} = k \qquad m - m' = p \qquad n - n' = q \qquad (7.15)$$

then $WVD_{h,h'}(t,\omega)$ in (7.14) have the same form as $H(t,\omega)$ in the 2D Gabor expansion (7.2). In this case, the harmonic frequencies, pT and $q\Omega$, are determined by the distance between $h_{m,n}(t)$ and $h_{m',n'}(t)$. The farther $h_{m,n}(t)$ and $h_{m',n'}(t)$ are apart, the higher the harmonic frequencies contained in $WVD_{h,h'}(t,\omega)$.

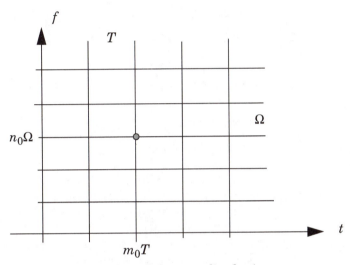

Fig. 7–2 Gabor sampling lattice

Fig. 7–3 illustrates the locations of $WVD_{h,h'}(t,\omega)$. Each ellipse is the superposition of the number of infinite cross Wigner-Ville distributions $WVD_{h,h'}(t,\omega)$ or energy atoms. For example, all $h_{m,n}(t)$ and $h_{m',n'}(t)$, in which $m_0 = (m + m')/2$ and $n_0 = (n + n')/2$, will create a $WVD_{h,h'}(t,\omega)$ that is centered at $(m_0T, n_0\Omega)$.

Note that (7.13) is not equivalent to (7.2). As shown by (7.15), the selections of $i, k, p,$ and q are not independent in (7.13). Unlike the case of 2D Gabor expansion, (7.13) implies that each ellipse in Fig. 7–3 only contains certain types of energy atoms. For example, the DC terms, $m = m'$ and $n = n'$, only appear in the Gabor sampling grids (shaded ellipses). The significance of (7.13), however, is that it can be directly computed by 1D Gabor expansion. This decomposition scheme not only saves substantial computations, but also is amendable for the discrete implementation, as we shall see soon.

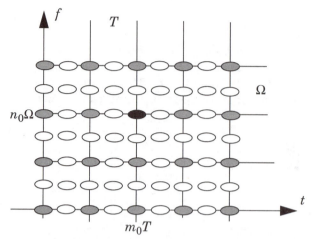

Fig. 7–3 Each ellipse represents the superposition of infinite *elementary Wigner-Ville distribution* (energy atom). All $h_{m,n}(t)$ and $h_{m',n'}(t)$, in which $(m + m')/2 = m_0$ and $(n + n')/2 = n_0$, will create a term centered at $(m_0 T, n_0 \Omega)$.

If the Wigner-Ville distribution is considered as the signal energy distribution in the joint time-frequency domain, then Fig. 7–3 says that the signal energy can be broken up into an infinite number of "molecules" (as indicated by ellipses). All those molecules are concentrated and *symmetrical*. Moreover, each molecule is the superposition of an infinite number of atoms. The rate of oscillation of atoms is different, but they all have the same 2D Gaussian envelope.

Based on the decomposition of the Wigner-Ville distribution in (7.13), we define the *time-frequency distribution series* (TFDS)[*] as

$$\text{TFDS}_D(t, \omega) = \sum_{d=0}^{D} P_d(t, \omega) \tag{7.16}$$

where $P_d(t,\omega)$ is the set of those $\text{WVD}_{h,h'}(t,\omega)$ which have a similar contribution to the useful properties and similar influence to the cross-terms. Because the impact to the cross-term as well as the useful properties are determined by the harmonic frequencies $|p|T = |m-m'|T$ and $|q|\Omega = |n-n'|\Omega$, $P_d(t,\omega)$ can also be considered as the set of $\text{WVD}_{h,h'}(t,\omega)$ in which $|m-m'| + |n-n'| = d$, i.e.,

$$P_d(t, \omega) = \sum_{|m-m'| + |n-n'| = d} C_{m,n} C_{m',n'} \text{WVD}_{h,h'}(t, \omega) \tag{7.17}$$

---

[*] Because it is the Gabor expansion and WVD-based time-dependent spectrum, the time-frequency distribution series is also known as *Gabor spectrogram* in the industry.

Substituting (7.14) into (7.17) yields

$$P_d(t, \omega) = 2 \sum_{|m - m'| + |n - n'| = d} C_{m,n} C^*_{m',n'} \exp\left\{ -j \frac{n + n'}{2} \Omega (m - m') T \right\} \qquad (7.18)$$

$$\exp\left\{ -\alpha \left(t - \frac{m + m'}{2} T \right)^2 - \frac{1}{\alpha} \left(\omega - \frac{n + n'}{2} \Omega \right)^2 \right\}$$

$$\exp\left\{ j[(m - m')T\omega + (n - n')\Omega t] \right\}$$

Obviously, the parameter $d = |m - m'| + |n - n'|$ reflects the rate of oscillation of $\mathrm{WVD}_{h,h'}(t,\omega)$. The parameter D in (7.16) denotes the order of the time-frequency distribution series. $\mathrm{TFDS}_D(t,\omega)$ contains up to Dth order $P_d(t,\omega)$.

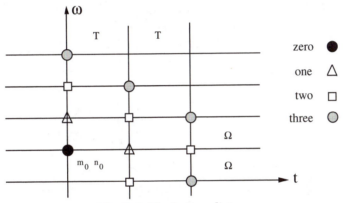

Fig. 7–4 Manhattan distance.

The parameter d can be remembered as the *Manhattan distance* between $h_{m,n}(t)$ and $h_{m',n'}(t)$. The Manhattan distance is the terminology commonly used in the contents of VLSI (very large scale integration) design, which defines the distance between two points as the vertical distance plus the horizontal distance. For example, the distance between $a_1(x1,y1)$ and $a_2(x2,y2)$ is $|x1-x2| + |y1-y2|$, as shown in Fig. 7–4, rather than the usual Euclidean distance,

$$\sqrt{(x1 - x2)^2 + (y1 - y2)^2}$$

Apparently, $\mathrm{TFDS}_0(t,\omega) = P_0(t,\omega)$, when $D = 0$. In this case,

$$\mathrm{TFDS}_0(t, \omega) = P_0(t, \omega) = 2 \sum_{m,n} |C_{m,n}|^2 \exp\left\{ -\alpha(t - mT)^2 - \frac{1}{\alpha}(\omega - n\Omega)^2 \right\} \qquad (7.19)$$

which is non-negative. The right side of (7.19) can be thought of as a 2D interpolation filter. As shown in (7.12), the Gabor coefficients $C_{m,n}$ are the sampled STFT with the analysis function $\gamma(t)$, and thereby the inputs of the 2D interpolation filter $|C_{m,n}|^2$ are the sampled STFT spectrogram. The impulse response of the 2D interpolation filter is a 2D Gaussian function that is optimally concentrated in the joint time-frequency domain. Consequently, the zero order time-frequency distribution series is similar to the STFT spectrogram with the analysis function $\gamma(t)$. As D goes to infinity, the $\text{TFDS}_\infty(t,\omega)$ manifestly converges to the Wigner-Ville distribution. By adjusting the order D, we can effectively balance the cross-term interference, useful properties, and resolution.

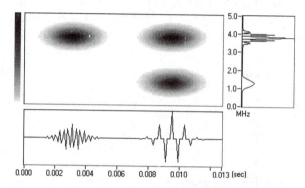

Fig. 7–5 $\text{TFDS}_0(t,\omega)$ is close to the STFT spectrogram.

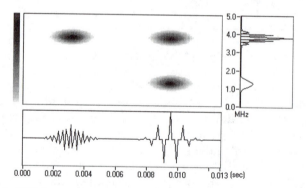

Fig. 7–6 $\text{TFDS}_3(t,\omega)$ has good resolution but no cross-term interference.

Fig. 7–5 to Fig. 7–7 illustrate the TFDS with different orders. When $D = 0$ in Fig. 7–5, there is no undesirable term, but the resolution is rather poor (similar to the STFT spectrogram). For $D = 3$ in Fig. 7–6, the resolution is close to the WVD, but there is no undesirable term. Moreover, it can be shown that $\text{TFDS}_3(t,\omega)$ well approximate the properties of the WVD. As the order gets larger, the unwanted term becomes visible and the TFDS converges to the WVD in Fig. 7–7. In general, the higher the order, the better the resolution. On the other hand, the higher the order, the severer the interference. A good compromise usually is between order two to four.

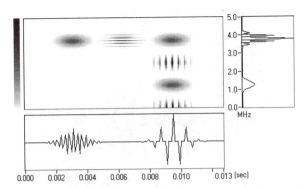

Fig. 7–7 TFDS$_{10}(t,\omega)$ converges to the WVD. (A high-order TFDS improves the resolution but also introduces interference. A good compromise has been found for order two to four.)

Fig. 7–8 and Fig. 7–9 compare the Wigner-Ville distribution, the Choi-Williams distribution, and TFDS$_3(t,\omega)$ for three Gaussian functions. Note that all the unwanted terms caused by the two components with the same time or frequency center remain in the CWD.

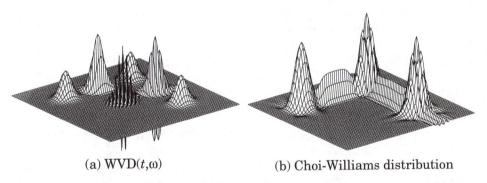

(a) WVD(t,ω) (b) Choi-Williams distribution

Fig. 7–8 The Choi-Williams distribution suppresses the cross-term created by two Gaussian functions with different time and frequency centers, but it preserves the cross-terms caused by two Gaussian functions with either the same time or same frequency centers.

Fig. 7–7 and Fig. 7–8 illustrate three cross-terms created by each pair of $h_{m,n}(t)$ and $h_{m',n'}(t)$. Traditionally, the cross-terms are always considered as undesirable and bad. Thereby they should simply be removed. Eq. (7.7) demonstrates, however, that the cross-term is not always unwanted. The significance of each WVD$_{h,h'}(t,\omega)$ depends on the Manhattan distance between those two auto-terms, $h_{m,n}(t)$ and $h_{m',n'}(t)$. When WVD$_{h,h'}(t,\omega)$ correspond to the pair of $h_{m,n}(t)$ and $h_{m',n'}(t)$ which are close to each other, they play important roles for the useful properties and therefore cannot be neglected, even though they are the "cross-terms."

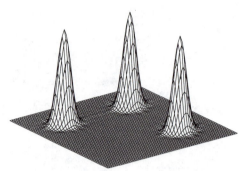

Fig. 7–9 The resolution of $\text{TFDS}_3(t,\omega)$ is close to the WVD, but there is no undesirable term. Moreover, it can be shown that $\text{TFDS}_3(t,\omega)$ well approximate the properties of the WVD.

As discussed in Chapter 2, a given signal can be decomposed in an infinite number of ways. Different decomposition schemes will lead to different cross-terms. In other words, the cross-term is not unique. The concept of cross-term in fact is ambiguous. It is the high-order harmonic terms that we should remove from the Wigner-Ville distribution rather than general cross-terms.

Finally, let's investigate how the time-frequency distribution series is related to the mean instantaneous frequency. In what follows, we use the instantaneous bandwidth as the measure to evaluate different representation schemes. The instantaneous bandwidth $\Delta_\omega(t)$ is defined by

$$\Delta_\omega^2(t) = \frac{\int (\omega - \langle\omega\rangle_t)^2 P(t, \omega)d\omega}{\int P(t, \omega)d\omega} \tag{7.20}$$

where $\langle\omega\rangle_t$ denotes the conditional mean frequency defined in (7.1). The instantaneous bandwidth $\Delta_\omega(t)$ is the indication of the signal energy spread with respect to the conditional mean frequency. Intuitively, the good time-dependent spectrum $P(t,\omega)$ should be such that its conditional mean frequency is equal to the signal mean instantaneous frequency, the first derivative of a signal's phase. The instantaneous bandwidth should be as small as possible.

Fig. 7–10 illustrates the narrow window STFT spectrogram and $\text{TFDS}_3(t,\omega)$ for the linear chirp signal. As mentioned earlier, the conditional mean frequency of the STFT spectrogram in general is not equal to the first derivative of the phase of the signal. When the narrow window is applied, however, the conditional mean frequency computed by the STFT spectrogram is very close to the first derivative of the phase. Fig. 7–11 depicts the theoretical mean instantaneous frequency (the first derivative of the phase) and conditional mean frequency computed by the STFT spectrogram and $\text{TFDS}_3(t,\omega)$. In this case, three lines (theoretical value, mean instantaneous frequencies computed by STFT spectrogram, and TFDS$_3(t,\omega)$) basically are indistinguishable. However, the instantaneous bandwidth of TFDS$_3(t,\omega)$ is much smaller than that of the STFT spectrogram as shown in Fig. 7–11. The same observation can also be obtained from the non-linear chirp signals in Fig. 7–12 and Fig. 7–13.

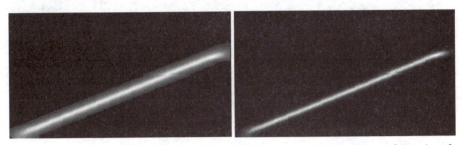

Fig. 7–10 STFT spectrogram (left) and TFDS$_3(t,\omega)$ (right) for linear chirp signal.

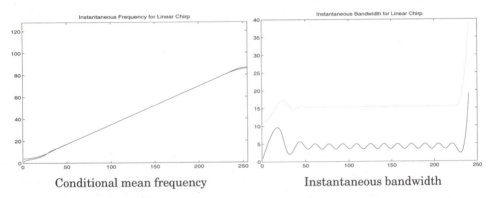

Conditional mean frequency   Instantaneous bandwidth

Fig. 7–11 The theoretical mean instantaneous frequency and the conditional mean frequency computed by the STFT spectrogram and the TFDS$_3(t,\omega)$ basically are indistinguishable. The energy spread of TFDS$_3(t,\omega)$ (solid line), however, is much smaller than that in the STFT spectrogram (dotted line).

As the numerical simulations indicated, the conditional mean frequency of the TFDS not only well approximates the signal's mean instantaneous frequency, but also has much smaller instantaneous bandwidth. Although the conditional mean frequency of the WVD is equal to the first derivative of the signal's phase, its instantaneous bandwidth widely oscillates for the non-linear chirp signal.

In addition to the STFT spectrogram, Wigner-Ville distribution, and time-frequency distribution series, there are more than a dozen other bilinear distributions which presumably preserve the properties of the Wigner-Ville distribution. However, because they all are developed from continuous-time Cohen's class, the properties derived in general do not hold for the discrete-time signals. On the other hand, the time-frequency distribution series, as shown in (7.18), has the closed form for arbitrary signals. Then, we could directly sample the continuous-time time-frequency distribution series to obtain the discrete time-frequency distribution series. Consequently, all the properties possessed by the continuous-time time-frequency distribution series will automatically be carried over to its discrete-time counterpart.

Fig. 7–12 STFT spectrogram (left) and TFDS$_3(t,\omega)$ (right) for non-linear chirp signal.

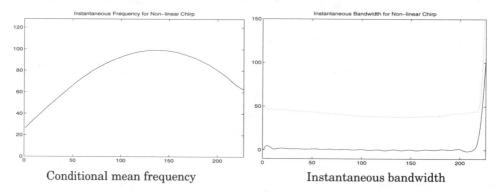

Conditional mean frequency   Instantaneous bandwidth

Fig. 7–13 The theoretical mean instantaneous frequency and the conditional mean frequency computed by the STFT spectrogram and the TFDS$_3(t,\omega)$ basically are indistinguishable. The energy spread of TFDS$_3(t,\omega)$ (solid line), however, is much smaller than that in the STFT spectrogram (dotted line).

7.3   Discrete Time-Frequency Distribution Series

In the preceding section, we discussed the *time-frequency distribution series* for the continuous-time signals. We showed that the low-order TFDS not only significantly reduces the unwanted oscillation, but also well depicts the signal's time-dependent spectrum. Another important feature of the TFDS is that the TFDS has a closed form for arbitrary signals $s(t)$. The only thing it needs to compute are the Gabor coefficients $C_{m,n}$. As discussed in Chapter 3, the Gabor coefficients are no more than the sampled short-time Fourier transform. And the algorithm of computing STFT is well established. Because of this nice feature, the discrete version of the time-frequency distribution series can be obtained by directly sampling its continuous-time counterpart.

For example, if the signal $s(t)$ is band limited, then the *discrete time-frequency distribution series* is defined as

$$\text{DTFDS}_D[i, k] \equiv \text{TFDS}_D(t, \omega)\Big|_{t = i\Delta t, \, \omega = \frac{2\pi k}{L\Delta t}} \qquad \text{for } -\frac{L}{2} \leq k < \frac{L}{2} \qquad (7.21)$$

where $1/\Delta t$ denotes the sampling frequency. L denotes the total number of frequency bins. Because the discrete time-frequency distribution series is the sampled version of the continuous-time time-frequency distribution series, the discrete time-frequency distribution series will automatically inherit (at least well approximate) the properties possessed by its continuous-time counterpart. There is no aliasing problem as happened in the Wigner-Ville distribution.

For the sake of presentation simplicity, let $\Delta t = 1$ in (7.21). Without introducing any confusion, we use $\text{TFDS}_D[i,k]$ for the discrete time-frequency distribution series. Then, the discrete time-frequency distribution series can be summarized as

$$\text{TFDS}_D[i, k] \equiv \sum_{d=0}^{D} P_d[i, k] \qquad (7.22)$$

where

$$P_d[i, k] \equiv \sum_{|m - m'| + |n - n'| = d} C_{m, n} C_{m', n'} \text{WVD}_{h, h'}[i, k] \qquad (7.23)$$

The $\text{TFDS}_D[i,k]$ in fact is the sum of all $\text{WVD}_{h,h'}[i,k]$ in which the Manhattan distance of the corresponding Gabor elementary functions $h_{m,n}[i]$ and $h_{m'n'}[i]$ is less than or equal to D. $\text{WVD}[i,k]$ is defined as a *sampled Wigner-Ville distribution*, i.e.,

$$\text{WVD}_s[i, k] = \text{WVD}_s(t, \omega)\Big|_{t = i\Delta t, \, \omega = \frac{2\pi k}{L\Delta t}} \qquad (7.24)$$

where Δt denotes the sampling interval. For the Gaussian functions, *WVD* is obtained by sampling the formula (7.14), i.e.,

$$\text{WVD}_{h, h'}[i, k] \equiv 2 \exp\left\{ -\alpha\left(i - \frac{m + m'}{2}\Delta M\right)^2 - \frac{1}{\alpha}\left(k - \frac{n + n'}{2}\Delta N\right)^2 \right\} \qquad (7.25)$$

$$\times \exp\left\{ j\frac{2\pi}{L}\left[(m - m')\Delta M k + (n - n')\Delta N i - \frac{n + n'}{2}\Delta N(m - m')\Delta M \right] \right\}$$

where we assume that $\Delta t = 1$. Note that $\text{WVD}_{h,h'}[i,k]$ in (7.25) is completely determined by the parameters of the Gabor expansion, such as ΔM, ΔN, L, and α, which are independent of the analyzed signal. Therefore, once ΔM, ΔN, L, and α are determined, we can precompute $\text{WVD}_{h,h'}[i,k]$ and save them in the table.

Consequently, the evaluation of the discrete time-frequency distribution series involves no more than looking up the table.

Because $\mathrm{WVD}_{h,h'}[i,k] = \mathrm{WVD}^*_{h',h}[i,k]$, $P_d[i,k]$ can be further simplified as

$$P_d[i,k] = Re\left\{ \sum_{(m-m')+|n-n'|=d} A(m,m',n,n') \right. \tag{7.26}$$

$$\times \exp\left[-\alpha\left(i - \frac{m+m'}{2}\Delta M\right)^2 - \frac{1}{\alpha}\left(\frac{2\pi}{L}\right)^2\left(k - \frac{n+n'}{2}\Delta N\right)^2\right]$$

$$\left. \times \exp\left[j\frac{2\pi}{L}[(m-m')\Delta Mk + (n-n')\Delta Ni]\right]\right\} \qquad m > m'$$

The weight $A(m,m',n,n')$ is equal to

$$\begin{cases} 2|C_{m,n}|^2 & m=m' \text{ and } n=n' \\[2ex] 4C_{m,n}C^*_{m',n'}\exp\left\{-j\frac{n+n'}{2}\Delta N(m-m')\Delta M\right\} & \text{otherwise} \end{cases}$$

Compared to (7.25), the formula (7.26) saves half the computations.

The computation complexity grows as the order D increases. In real applications, however, the order D is seldom larger than four. When D is bigger than four, unwanted terms will become noticeable.

7.4   Selections of $h[i]$ and $\gamma[i]$

Based on the previous discussions, it is obvious that the convergence of the time-frequency distribution series is independent of the selection of the Gabor elementary function $h[i]$ and the dual function $\gamma[i]$. As the order D goes to infinity, the time-frequency distribution series always converges to the Wigner-Ville distribution, no matter how we choose $h[i]$ and $\gamma[i]$. To avoid unwanted terms, however, we would like to keep the order of time-frequency distribution series as low as possible. This implies that zeroth order TFDS, $\mathrm{TFDS}_0[i,k]$, should be as close as possible to the true time-dependent spectrum. If this is the case, we could obtain the desired time-dependent spectrum by adding a few lower order harmonics. The resulting presentation not only well delineates the signal's time-dependent spectrum, but also has limited interference. Otherwise, we would have to include a considerable number of high-order terms to obtain the meaningful time-dependent spectrum. In this case, the results will be similar to the WVD. To better balance the unwanted terms and useful properties, it is desired that:

- *The analysis function $\gamma[i]$ is localized in both time and frequency domains;*
- *$h[i]$ and $\gamma[i]$ have similar time/frequency centers and time/frequency resolutions.*

These concepts may not be obvious at first glance. In order to get a better understanding, let's look at a couple of examples. First, let's examine the zeroth order time-frequency distribution series, e.g.,

$$\text{TFDS}_0[i, k] = 2 \sum_{m=-\infty}^{\infty} \sum_{n=-\infty}^{\infty} |C_{m,n}|^2 \exp\left\{-\alpha(i - m\Delta M)^2 - \frac{1}{\alpha}\left(\frac{2\pi}{L}\right)^2 (k - n\Delta N)^2\right\}$$

Notice that the exponential parts are the Wigner-Ville distributions of $h_{m,n}[i]$, which are concentrated at $(m\Delta M, n\Delta N)$. The weight $C_{m,n}$ is the projection of signal $s[i]$ on $\gamma_{m,n}[i]$, that is, $C_{m,n} = \langle s, \gamma_{m,n}\rangle$. If $\gamma_{m,n}[i]$ is substantially different from $h_{m,n}[i]$, for instance, $\gamma_{m,n}[i]$ is not centered at $(m\Delta M, n\Delta N)$, then $|C_{m,n}|^2$ will not reflect the signal's behavior in the vicinity of $(m\Delta M, n\Delta N)$.

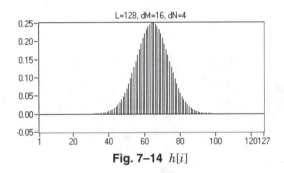

Fig. 7–14 $h[i]$

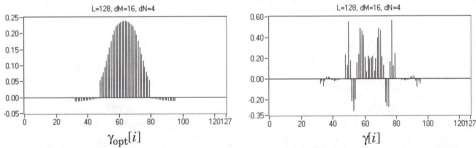

$\gamma_{opt}[i]$ $\gamma[i]$

Fig. 7–15 Although both dual functions lead to perfect reconstruction, their shapes are dramatically different. $\gamma_{opt}[i]$ is optimally close to $h[i]$. Therefore, it has similar time and frequency centers to those of $h[i]$. $\gamma_{opt}[i]$ is optimally concentrated in the joint time-frequency domain, whereas the right-side dual function is neither concentrated in time nor in frequency.

Fig. 7–14 and Fig. 7–15 depict the Gabor elementary function $h[i]$ and corresponding dual functions $\gamma[i]$. Although both dual functions in Fig. 7–15 lead to perfect reconstruction, their shapes are dramatically different. $\gamma_{opt}[i]$ is optimally close to $h[i]$. Therefore, it has similar time and frequency centers as those of $h[i]$. Because $h[i]$ is a Gaussian-type function, which is optimally concentrated in the joint time-frequency domain, $\gamma_{opt}[i]$ is also localized. On the other hand, the dual function depicted on the right side of Fig. 7–15 is neither concentrated in time nor in frequency[*].

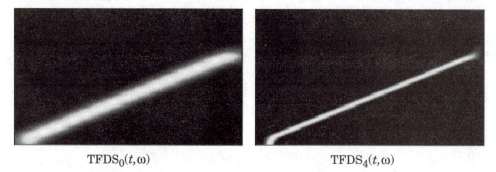

TFDS$_0(t,\omega)$ TFDS$_4(t,\omega)$

Fig. 7–16   Time-frequency distribution series with optimal dual function $\gamma_{opt}[k]$. (Because of the good initial guess, TFDS$_0(t,\omega)$, we obtain the desired representation by adding a few low-order harmonic terms.)

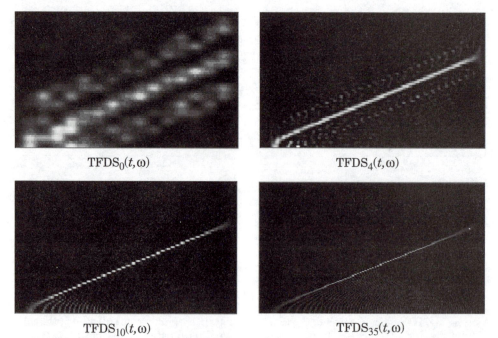

TFDS$_0(t,\omega)$ TFDS$_4(t,\omega)$

TFDS$_{10}(t,\omega)$ TFDS$_{35}(t,\omega)$

Fig. 7–17   Time-frequency distribution series with bad dual function. (Although it converges to the WVD, the lower order TFDS is way off the true time-dependent spectra.)

Fig. 7–16 and Fig. 7–17 illustrate the TFDS, computed by the two different dual functions plotted in Fig. 7–15, for the linear chirp signal whose frequencies

---

[*] The algorithm of computing $\gamma[k]$ that is intentionally different from $h[k]$ is introduced in Appendix B.

linearly increases with respect to time. In Fig. 7–16, because $\gamma_{opt}[i]$ has the same time and frequency center as $h[i]$ and is localized, the low-order TFDS well delineates the linear chirp signal with negligible interference. Fig. 7–17 plots the results of TFDS for the same chirp signal but with dual function $\gamma[i]$ (plotted as shown in the right side of Fig. 7–15), whose shape considerably differs from $h[i]$. Although in this case $\text{TFDS}_{\infty}[i,k] \rightarrow \text{DWVD}[i,k]$, the low-order TFDS does not reflect the true time-dependent spectrum. The meaningful linear chirp pattern is not obtained until the order increases to ten. At the same time, $\text{TFDS}_{10}(t,\omega)$ has included substantial interference, as shown. Therefore, to achieve the desired joint time-frequency representation, $\gamma[i]$ not only has to be localized, but also should be close to $h[i]$. The solution is the *orthogonal-like Gabor expansion* introduced in Chapter 3.

The selection of the variance α that controls the time and frequency resolution of the Gabor elementary function also affects the performance of the lower order time-frequency distribution series, though its impact is much smaller than that in the STFT spectrogram. A good choice depends on the applications at hand. The principle of the selection, however, is similar to the selection of the window function for the short-time Fourier transform. If time resolution is important, for instance, when we try to catch a short duration pulse, then the larger α is favored. Otherwise, we should use smaller α, which leads to better frequency resolution.

7.5 Time-Frequency Distribution Series and Cohen's Class

The best way to understand the difference between the time-frequency distribution series and Cohen's class is from the ambiguity domain. Let's take the ambiguity transformation, with respect to the Gabor expansion in (7.10), yields

$$\text{AF}(\vartheta, \tau) = \sum_{m,n} \sum_{m',n'} C_{m,n} C_{m',n'} \text{AF}_{h,h'}(\vartheta, \tau) \tag{7.27}$$

As discussed in Chapter 6, the partial derivative of the phase of $\text{WVD}_{h,h'}(t,\omega)$ is equal to the center of $\text{AF}_{h,h'}(\vartheta,\tau)$ (see (6.17)). Therefore, retaining the low-order terms of time-frequency distribution series (low harmonics or small partial derivative of the phase) is equivalent to keeping those $\text{AF}_{h,h'}(\vartheta,\tau)$ which are centered in the vicinity of the origin. The resulting time-dependent auto-correlation is

$$R(t, \tau) = \frac{1}{\sqrt{2\pi}} \int \Im\{\text{AF}(\vartheta, \tau)\} e^{jt\vartheta} d\vartheta \tag{7.28}$$

where $\Im\{*\}$ denotes a *non-linear truncation*. If taking the Fourier transformation with respect to (7.28), we will obtain a truncated time-frequency distribution series. On the other hand, as long as the kernel function $\phi(\vartheta,\tau)$ is continuous for all ϑ and τ, the Cohen's time-dependent auto-correlation is a typical linear filter, i.e.,

$$R(t, \tau) = \frac{1}{\sqrt{2\pi}}\int\Phi(\vartheta, \tau)AF(\vartheta, \tau)e^{jt\vartheta}d\vartheta$$

As discussed in Chapter 6, $AF_{h,h'}(\vartheta,\tau)$ has the Gaussian envelope that extends to the entire ambiguity domain. In other words, $AF_{h,h'}(\vartheta,\tau)$ are heavily overlapped. Therefore, there is no closed form of the kernel function $\Phi(\vartheta,\tau)$ that is equivalent to truncating the $AF(\vartheta,\tau)$ in (7.27).

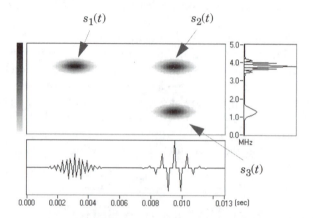

Fig. 7–18 Three-tone test signal.

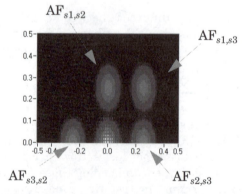

Fig. 7–19 $|AF(\vartheta,\tau)|$ of three Gaussian functions. (Except for the one that is concentrated in the origin, all other clusters correspond to the cross-terms.)

As we have noticed, many well-known members of Cohen's class, such as the Choi-Williams distribution [23], emphasize preserving the properties of the Wigner-Ville distribution. In order to "precisely" keep the time/frequency marginal conditions, the kernel has to be

$$\Phi(\vartheta, 0) = \Phi(0, \tau) = 1 \qquad \forall\, \vartheta, \tau \tag{7.29}$$

In other words, all $AF(\vartheta,\tau)$ that are close to the ϑ-axis or τ-axis will be preserved, no matter how far they are from the origin (0,0) or how important they are to the useful properties. On the other hand, $AF(\vartheta,\tau)$ concentrated at the τ-axis correspond to the cross-terms caused by two signals with the same frequency center. $AF(\vartheta,\tau)$ concentrated at the ϑ-axis correspond to the cross-term caused by two signals with the same time center. The direct consequence of using the kernel $\Phi(\vartheta,\tau)$ described in (7.29) is that all cross-terms caused by two signals that have either the same time or frequency centers will be present.

Fig. 7–19 illustrates the amplitude of the ambiguity function of three Gaussian functions. The corresponding Wigner-Ville distribution is plotted in Fig. 7–18. Note that, except for the one that is concentrated in the origin, all other clusters correspond to the cross-terms. The one that is centered in the τ-axis, $AF_{s1,s2}$, corresponds to that caused by two Gaussian functions having the same frequency center. Two centered in the ϑ-axis, $AF_{s2,s3}$ and $AF_{s3,s2}$, correspond to the cross-AF caused by two Gaussian functions having the same time center. The one away from both axes, $AF_{s1,s3}$, corresponds to the cross-term caused by two Gaussian functions that have different time as well as frequency centers.

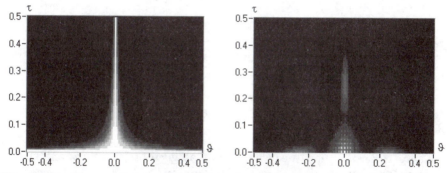

Fig. 7–20   Exponential kernel (left) and corresponding $|\Phi(\vartheta,\tau)AF(\vartheta,\tau)|$ (right). ($|\Phi(\vartheta,\tau)AF(\vartheta,\tau)|$ contains all portions of AF that are on both the ϑ-axis and τ-axis.)

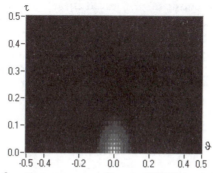

Fig. 7–21   Ambiguity function corresponding to $TFDS_3(t,\omega)$. (The auto-terms are well preserved but all cross-terms are removed.)

Fig. 7–20 illustrates the exponential kernel function

$$\Phi(\vartheta, \tau) = \exp\{-\alpha(\vartheta\tau)^2\}$$

which was first proposed by Choi and Williams [23] and extensively studied by many others [140]. Because the exponential kernel satisfies (7.29), the product $|\Phi(\vartheta,\tau)AF(\vartheta,\tau)|$ contains all cross-terms that are either in the ϑ-axis or τ-axis.

Fig. 7–21 depicts the corresponding ambiguity function to $\text{TFDS}_3(t,\omega)$. The comparison of the Choi-Williams distribution and the low order time-frequency distribution series is plotted in Fig. 7–22. Although in $\text{TFDS}_3(t,\omega)$ the high harmonic terms, starting at four, are removed, $\text{TFDS}_3(t,\omega)$ still well approximates the marginal as well as other properties. This is because the high harmonics have very limited contributions to the useful properties.

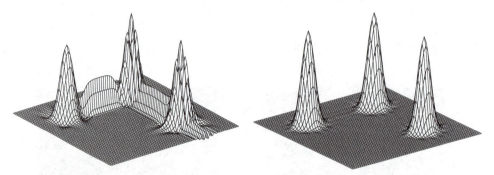

Fig. 7–22 $\text{TFDS}_3(t,\omega)$ (right) better delineates the time-dependent spectra than the Choi-Williams distribution (left).

There is an attitude these days that a "good" time-dependent spectrum must *completely* satisfy all the properties listed in Table 6–1 . To do so, one has to preserve all portions on both the ϑ-axis and τ-axis, even if they are far from the origin. The direct consequences are that the resulting presentations would contain a tremendous amount of unwanted interference. It is clear now that to reduce undesirable interference and to preserve some properties, such as the time marginal condition, the frequency marginal condition, etc., are two conflicting objectives. If we cannot accommodate those two goals simultaneously, which one is more important? The low-order time-frequency distribution possesses a very limited interference. Although it does not completely satisfy the properties listed in Table 6–1 , the differences are small, which could be neglected for most real applications.

To retain those $AF_{h,h'}(\vartheta,\tau)$ which are centered in the vicinity of the origin is somehow similar to applying a kernel function $\Phi(\vartheta,\tau)$, defined as

$$\Phi(\vartheta,\tau) = \begin{cases} 1 & \left(\dfrac{\vartheta}{a}\right)^2 + \left(\dfrac{\tau}{b}\right)^2 \leq r \\ 0 & \text{otherwise} \end{cases} \tag{7.30}$$

The parameters a and b determine the range of the kernel in two axes. The small radian r corresponds to the low-order harmonics. As the r increases, high harmonics are included. As r goes to infinity, the resulting time-dependent power spectrum approaches the Wigner-Ville distribution. However, because of the sharp cut in the edge of

$$\frac{\vartheta^2}{a^2} + \frac{\tau^2}{b^2} = r$$

the resulting time-dependent power spectrum (double Fourier transform of the ambiguity function) will have strong ripples. To overcome this problem, Wu and Morris investigated the kernel with smoothing transition [191].

The results obtained by Wu and Morris have many similarities with those obtained by the time-frequency distribution series. The fundamental difference between the members of Cohen's class and the time-frequency distribution series is that Cohen's class represents the linear filtering, and truncating the time-frequency distribution series is typical non-linear processing.

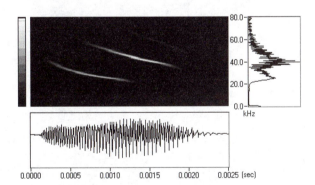

Fig. 7–23 Bat sound computed by the fourth-order TFDS. (Bat data are provided by Curtis Condon, Ken White, and Al Feng of the Beckman Institute at the University of Illinois.)

Before finishing this chapter, let's compare all the different time-dependent spectra, introduced in this book, for a bat sound. Fig. 7–23 plots the result computed by the fourth-order time-frequency distribution series. The echo-location pulse was emitted by a large brown bat, *Eptesicus fuscus*. Fig. 7–24 depicts results obtained by other algorithms introduced in the previous chapters. The STFT spectrogram is non-negative, but it has the lowest resolution. The TFDS and signal-dependent kernel distribution have better resolution without the significant presence of high oscillations. The Choi-Williams distribution possesses strong horizontal as well as vertical ripples, whereas the cone-shape distribution only has relatively weaker ripples along the time index.

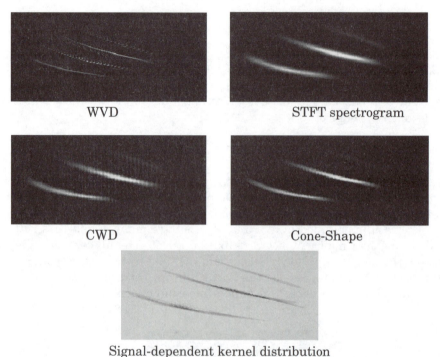

WVD   STFT spectrogram

CWD   Cone-Shape

Signal-dependent kernel distribution

Fig. 7–24 Bat sound computed by different time-frequency transforms

Summary

Using the *orthogonal-like Gabor expansion,* we can decompose the Wigner-Ville distribution into the 2D localized harmonic series known as the *time-frequency distribution series.* The useful properties of the Wigner-Ville distribution are obtained by *averaging* the Wigner-Ville distribution and they are mainly determined by the low-order harmonic terms. The high harmonic terms have very limited contribution to the useful properties, but are directly responsible for the cross-term interference. Selectively removing the high harmonics enables us to balance the useful properties and the interference. A lower order time-frequency distribution series not only well delineates the signal time-dependent spectra, but also significantly diminishes those annoying oscillations appearing in the Wigner-Ville distribution. Discarding the high harmonic terms is equivalent to suppressing the part of ambiguity functions that is far apart from the origin in the ambiguity domain. In contrast to Cohen's class, which is linear filtering on the ambiguity domain, truncating the time-frequency distribution series is a non-linear low-pass operation.

By using the time-frequency distribution series, we can gain a better understanding of the nature of the cross-terms. The phrase "cross-term" in fact is rather ambiguous. First, the cross-term is not unique; it is different for different decomposition schemes. Second, the cross-term is not always unimportant. When the cross-term is caused by a pair of auto-terms that are close to each other, it appears the lower order harmonics and it will play an important role for the useful properties. *It is the high harmonic terms that are undesirable for joint time-frequency representations.*

Adaptive Representation and Adaptive Spectrogram

*I*n previous chapters, we introduced the Gabor transform (or sampled short-time Fourier transform), wavelets, and several bilinear transformations. One primary motivation for these different schemes is to improve the joint time-frequency resolution with the least amount of cross-term interferences. As is often the case, time resolution and frequency resolution are in conflict. For example, in the case of the Gabor transform, to achieve a better time resolution, we chose a short duration window. But the shorter window reduces the frequency resolution. A long window will improve frequency resolution, but reduce time resolution. The window selection is a nontrivial problem in the Gabor transform and is often at the expense of sacrificing one property or the other.

On the other hand, the Wigner-Ville distribution does not suffer from the window problem, which possesses the best joint time-frequency resolution. However, the WVD has the problem of cross-term interference, which greatly limits its applications. The time-frequency distribution series (TFDS) is the decomposition of the Wigner-Ville distribution. Selectively removing the high harmonic terms leads to the high resolution representation with limited interference. The TFDS in general is rather robust. But for the situation where both very short duration and long duration signals are present, the effectiveness of the TFDS is also limited.

Fig. 8–1 and Fig. 8–2 show a synthetic signal where two signals are superimposed. One is a sinusoidal waveform and the other one is an impulse. Fig. 8–

1a and Fig. 8–1b show the STFT spectrogram (square of STFT) computed by a 16-point Hanning window and a 64-point Hanning window, respectively. The shorter window, Fig. 8–1a, reveals the impulse as indicated by the vertical line in the middle of the picture. The wide horizontal stripe indicates very poor frequency resolution. The longer window, Fig. 8–1b, clearly improves the frequency resolution at the expense of the time resolution. By STFT, it is difficult to achieve good time and frequency resolutions simultaneously.

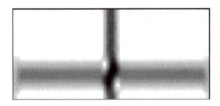
(a) by 16-point Hanning

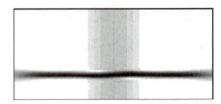
(b) by 64-point Hanning

Fig. 8–1 STFT spectrogram.

Fig. 8–2 shows the fourth-order TFDS, in which the wideband elementary function is used in order to catch a very short time-domain pulse. Compared with the STFT spectrogram, the TFDS holds up quite well in both the time and the frequency resolutions, though the low-order TFDS is also subject to the selection of elementary functions.

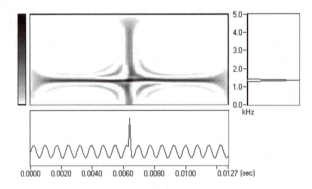

Fig. 8–2 $\text{TFDS}_4(t,\omega)$.

The desire for good resolution in both time and frequency domains is not artificial, but occurs in many real applications. Fig. 8–3 illustrates a simple radar target of a strip containing an open cavity[*]. A fundamental interest in the radar community is the nature of reflected electromagnetic signals. Based on the scattering physics of this structure, we anticipate there to be three scattering centers, corresponding to the left and right edge of the strip and the cavity exterior. Therefore, the reflected signal must contain three wideband time pulses

---

[*] A detailed description is presented in Section 10.2.

that are caused by these scattering centers. Besides these three scattering centers, we also expect to see some narrowband signals that correspond to the energy coupled into and re-radiated from the cavity. Those portions of the signal contain information on the resonant frequencies of the cavity. One fundamental task in radar image processing is to distinguish these two types of signals, those wideband scattering centers which can be used to form spatially meaningful radar images and those narrowband signals which convey target resonance information.

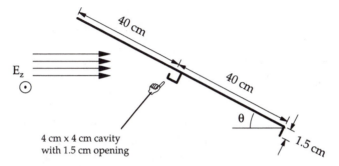

Fig. 8–3 Conducting strip with small cavity and fin

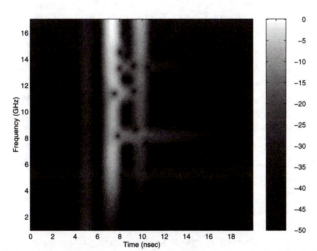

Fig. 8–4 STFT spectrogram of reflected signals in Fig. 8–3.

Fig. 8–4 displays an STFT spectrogram of the reflected electromagnetic signal. As expected, three scattering centers can be identified as vertical lines (wideband signals) in the STFT spectrogram. In addition to these scattering centers, we also see horizontal lines (narrowband signals), which correspond to the energy coupled into the cavity and re-radiated at its corresponding resonances. Although we gained some insights with the STFT spectrogram, the frequency resolution came at the cost of losing time resolution, and vice versa. Because the quality of radar imagery and the accuracy of target identification algorithms are closely related to the sharpness of both time and frequency resolutions, the Fourier transform-based spectrogram has been found to be inadequate in radar imaging applications.

Naturally, one would like to ask: Can we devise a system that is able to accommodate both narrowband and wideband signals? If we can, how do we make the window function adapt? These are the questions that will be addressed in this chapter.

In Section 8.1, we briefly introduce the general adaptive representation and adaptive spectrogram. Although the motivation for the adaptive representation was joint time-frequency analysis, the scheme of the adaptive representation turns out to be rather general. This means that not only can we essentially use an arbitrary function as the elementary function to match the analyzed signal, but we can also adjust any aspects of the elementary signals. For example, in addition to adapting the elementary function's duration, we can also tune up the frequency changing rate of the elementary functions.

In Section 8.2, we discuss the selection of the elementary function from the joint time-frequency analysis point of view. Because the Gaussian function is optimally concentrated in the joint time-frequency domain, it is natural to use the Gaussian function as the elementary function. The resulting adaptive representation is called the *adaptive Gabor representation*, because it largely resembles the conventional Gabor expansion. The Gaussian function-based adaptive spectrogram is non-negative and cross-term interference free.

One main issue of the adaptive representation is how to find the elementary function that best fits the analyzed signal. Section 8.3 is devoted to the algorithm of computing the optimal elementary functions. In particular, we discuss a heretical approach of selecting the best match between a signal and all the elementary functions. This approach has been implemented by several authors with a great deal of success.

Finally, in Section 8.4, the performance of the adaptive Gabor representation and other bilinear transformations are compared. We give some guidelines as to the expected performance of the adaptive spectrogram. For readers who are interested in the convergence problem of the adaptive expansion, a proof is given in the appendix. In this proof, it shows that, given enough elementary functions, the adaptive signal expansion indeed converges.

The adaptive Gabor representation and adaptive spectrogram were independently proposed by the authors (see [146] and [151]) and Mallat and Zhang [122] at about the same time. The adaptive representation initially was motivated by joint time-frequency analysis. Hence, the elementary functions had mainly been limited to the Gaussian-type functions. As a matter of fact, the selection of the elementary functions is much broader. We could almost use any functions as the elementary functions to achieve the adaptive representation. The adaptive representation introduced in this chapter is not only powerful for time-frequency analysis, but could also be used for many other applications, such as data compression and the de-noise process.

8.1 Adaptive Expansion and Adaptive Spectrogram

Analogous to the conventional Gabor expansion, we define the adaptive signal expansion as

$$s(t) = \sum_p B_p h_p(t) \tag{8.1}$$

where the coefficients are determined by

$$B_p = \langle s, h_p \rangle \tag{8.2}$$

which reflects the similarity between the signal $s(t)$ and the functions $h_p(t)$. Fig. 8–5 illustrates the procedure of the adaptive signal decomposition.

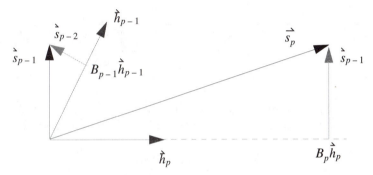

Fig. 8–5 Adaptive signal decomposition.

First, let's start with $p = 0$ and $s_0(t) = s(t)$, which is the original signal. Then, find the $h_0(t)$, among the set of the desired elementary functions, which is most similar to $s_0(t)$, in the sense of

$$|B_p|^2 = \max_{h_p} |\langle s_p(t), h_p(t) \rangle|^2 \tag{8.3}$$

for $p = 0$. The next step is to compute the residual, $s_1(t)$ by

$$s_{p+1}(t) = s_p(t) - B_p h_p(t) \tag{8.4}$$

Without loss of generality, let $h_p(t)$ have a unit energy. That is,

$$\|h_p(t)\|^2 = 1 \tag{8.5}$$

Then, the energy contained in the residual is

$$\|s_{p+1}(t)\|^2 = \|s_p(t)\|^2 - |B_p|^2 \tag{8.6}$$

Repeat (8.3) to find $h_1(t)$ that matches $s_1(t)$ the best, and so on. In each step, we find one elementary function, $h_p(t)$, that has the best match with $s_p(t)$. The main task of the adaptive signal expansion is to find a set of elementary functions, $\{h_p(t)\}$, that most resemble the signal's time-frequency structures, and in the meantime satisfy the formulae (8.1) and (8.2). Table 8–1 depicts the operations up - to pth steps.

Table 8–1

Residual	Projection	Energy in Residual
$s_0(t) = s(t)$	$B_0 = <s_0(t), h_0(t)>$	$\|s_0(t)\|^2$
$s_1(t) = s_0(t)\text{-}B_0h_0(t)$	$B_1 = <s_1(t), h_1(t)>$	$\|s_1(t)\|^2 = \|s_0(t)\|^2 - |B_0|^2$
... ...		
$s_p(t) = s_{p\text{-}1}(t)\text{-}B_{p\text{-}1}h_{p\text{-}1}(t)$	$B_p = <s_p(t), h_p(t)>$	$\|s_p(t)\|^2 = \|s_{p-1}(t)\|^2 - |B_{p-1}|^2$
... ...		

Intuitively, the residual will vanish if we continue to carry on the decompositions described by (8.3) and (8.4). To see this, let θ_p be the angle between $s_p(t)$ and $h_p(t)$, then

$$\cos\theta_p = \frac{\langle s_p, h_p \rangle}{\|s_p\|} = \frac{B_p}{\|s_p\|} \tag{8.7}$$

Applying the relationship (8.7) to (8.6) yields

$$\|s_p(t)\|^2 = \|s_{p-1}(t)\|^2 (\sin\theta_{p-1})^2 \tag{8.8}$$

Similarly, $\|s_{p-1}(t)\|^2$ can be written in terms of $s_{p-2}(t)$. If continuing this process, finally we have

$$\|s_p(t)\|^2 = \|s_0(t)\|^2 \prod_{i=0}^{p-1} (\sin\theta_i)^2 \leq \|s_0(t)\|^2 (\sin\theta_{max})^{2p} \tag{8.9}$$

where

$$|\sin\theta_{max}| = \max_{\theta_p} |\sin\theta_p| \qquad \forall p \tag{8.10}$$

Assume that at each step p, we are always able to find the optimal $h_p(t)$ that is not perpendicular to $s_p(t)$, that is,

$$\cos\theta_p \neq 0 \qquad \text{or} \qquad |\sin\theta_p| < 1 \qquad \forall\, p \tag{8.11}$$

then,

$$|\sin\theta_{max}| < 1 \tag{8.12}$$

Substituting (8.12) into (8.9) yields

$$\left\| s_p(t) \right\|^2 \leq \left\| s_0(t) \right\|^2 (\sin\theta_{max})^{2p} \to 0 \qquad as \qquad p \to \infty \tag{8.13}$$

which says that the energy of the residual signal exponentially decays and converges to zero[*]. In other words, we can perfectly reconstruct the signal $s(t)$ based on the set of adaptive elementary functions $\{h_p(t)\}$.

Obviously, for a different signal, its set of adaptive elementary $\{h_p(t)\}$ is also different. Each set of adaptive elementary functions $\{h_p(t)\}$ only works for one particular signal, not for all signals in L^2. Therefore, unlike the regular Gabor expansion (as well as wavelets), the set $\{h_p(t)\}$ constructed above will never be complete in L^2, even if the residual converges to zero.

Now, let's rewrite (8.6), such as

$$\left\| s_p(t) \right\|^2 = \left\| s_{p+1}(t) \right\|^2 + \left| B_p \right|^2 \tag{8.14}$$

which shows that the signal energy residual at the pth stage could be determined by the signal residual at $p+1$th stages plus B_p. Continuing to carry out this process, then we have

$$\left\| s(t) \right\|^2 = \sum_{p=0}^{\infty} \left| B_p \right|^2 \tag{8.15}$$

which is the energy conservation equation, and is similar to Parseval's relation in the Fourier transform.

Apply the Wigner-Ville distribution to both sides of (8.1) and arrange the terms in two groups

$$WVD_s(t, \omega) = \sum_p B_p^2 WVD_{h_p}(t, \omega) + \sum_{p \neq q} B_p B_q^* WVD_{h_p h_q}(t, \omega) \tag{8.16}$$

The first group represents the auto-terms and the second group represents the cross-terms. Since

$$\frac{1}{2\pi} \iint WVD_{h_p}(t, \omega) dt d\omega = \left\| h_p(t) \right\|^2 = 1 \tag{8.17}$$

---

[*] The convergent properties for the general cases are discussed in the appendix.

we have

$$\|s(t)\|^2 = \frac{1}{2\pi}\iint \mathrm{WVD}_s(t, \omega)dt d\omega$$

$$= \sum_p |B_p|^2 \frac{1}{2\pi}\iint \mathrm{WVD}_{h_p}(t, \omega)dt d\omega + \frac{1}{2\pi}\iint \sum_{p \neq q} B_p B_q{}^* \mathrm{WVD}_{h_p h_q} dt d\omega$$

$$= \sum_p |B_p|^2 + \frac{1}{2\pi}\iint \sum_{p \neq q} B_p B_q{}^* \mathrm{WVD}_{h_p h_q} dt d\omega \qquad (8.18)$$

Because of the relation described in (8.15), it is obvious that

$$\frac{1}{2\pi}\iint \sum_{p \neq q} B_p B_q{}^* \mathrm{WVD}_{h_p h_q} dt d\omega = 0 \qquad (8.19)$$

which implies that the second term in (8.18) contains zero energy. This gives us the reason to define a new time-dependent spectrum as

$$\mathrm{AS}(t, \omega) = \sum_p |B_p|^2 \mathrm{WVD}_{h_p}(t, \omega) \qquad (8.20)$$

Because it is an adaptive representation-based time-dependent spectrum, we call it an *adaptive spectrogram* (AS). Clearly, the adaptive spectrogram does not contain the cross-term interference, as occurred in the Wigner-Ville distribution, and it also satisfies the energy conservation relation

$$\|s(t)\|^2 = \frac{1}{2\pi}\iint \mathrm{AS}(t, \omega)dt d\omega \qquad (8.21)$$

As mentioned in the very beginning of the book, a fundamental issue of linear representation (or expansion) is the selection of the elementary functions. For the Gabor expansion, the set of elementary functions is made up of a time-shifted and frequency-modulated single prototype window function $h(t)$. In the wavelets, the elementary functions are obtained by dilating and translating a mother wavelet $\psi(t)$. In those two cases, the structures of elementary functions are simple and they can be determined in advance. The elementary functions employed for the adaptive representation, however, are relatively complicated. Usually, we first select a parametric model based on the applications at hand. Then, we continuously update the function $h_p(t)$. The better performance of the adaptive scheme is achieved at the cost of the matching process.

Finally, it is worthwhile to note that both the adaptive algorithm as well as adaptive spectrogram introduced in this section are independent of the selection of the elementary functions $h_p(t)$. Any functions can be used for the basic parametric model. Both algorithms hold for arbitrary elementary functions. In the next section, we shall discuss the selection of the elementary functions from the time-frequency analysis point of view.

8.2 Adaptive Gabor Representation (AGR)

In principle, the elementary functions used for the adaptive signal expansion of (8.1) can be very general. Practically, it may not be the case. To better characterize a signal's time-varying nature, it is desirable for the elementary functions to be localized in time and frequency simultaneously. Moreover, the elementary functions should be such that the resulting optimization algorithm is relatively easy to be implemented. Because the Gaussian-type signal achieves the lower bound of the uncertainty inequality, it is a natural selection for adaptive representation, that is,

$$h_p(t) = \left(\frac{\alpha_p}{\pi}\right)^{1/4} \exp\left\{-\frac{\alpha_p}{2}(t - T_p)^2\right\} \exp\{j\Omega_p t\} \tag{8.22}$$

where (T_p, Ω_p) is the time-frequency center of the elementary function. α_p^{-1} is the variance of the Gaussian function at (T_p, Ω_p).

The variance used in the regular Gabor expansion is fixed, while it is adjustable in (8.22). The Gaussian functions used in the regular Gabor expansions are located at fixed time and frequency grid points $(mT, n\Omega)$, while the centers of the elementary functions in (8.22) are not fixed and they can be anywhere. Adjusting the variance value will increase or decrease the duration of the elementary function; adjusting the parameters (T_p, Ω_p) will change the time and frequency centers of the elementary function. Thus, the overall effect of adjusting the variance and the time-frequency centers will allow us to better match the signal $s(t)$'s local time-frequency feature.

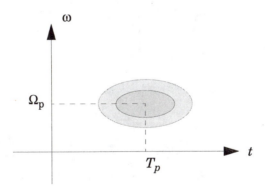

Fig. 8–6 The WVD of adaptive Gaussian functions.

Fig. 8–6 plots the joint time-frequency density function of the adaptive Gaussian functions, where

$$\text{WVD}_{h_p}(t, \omega) = 2\exp\left\{-\left[\alpha_p(t - T_p)^2 + \frac{(\omega - \Omega_p)^2}{\alpha_p}\right]\right\} \tag{8.23}$$

which is computed by the Wigner-Ville distribution. The joint time-frequency density function of the adaptive Gaussian functions is an ellipse and centered at (T_p, Ω_p). As introduced in Chapter 5, the energy concentration of the Gaussian-type functions is optimal. By using the Gaussian functions of different variance and different time-frequency center, each Gaussian function will then character-ize the local behavior of the signal, $s(t)$, that is being analyzed. If the signal at a particular instant of time exhibits an abrupt change in time, then a Gaussian function with a very small variance value may be used to match the abrupt change. If the signal exhibits a stable frequency for a long time, a Gaussian func-tion with a large variance value may be used.

Substituting (8.22) into (8.2) yields

$$s(t) = \sum_p B_p h_p(t) = \sum_p B_p \left(\frac{\alpha_p}{\pi}\right)^{1/4} \exp\left\{-\frac{\alpha_p}{2}(t-T_p)^2\right\} \exp\{j\Omega_p t\} \qquad (8.24)$$

which is called the adaptive Gabor representation (AGR), because it largely resembles the regular Gabor expansion.

Unlike the Gabor expansion, where the analysis and synthesis functions in general are not identical, the adaptive representation has the same analysis and synthesis functions. Once we have obtained the optimal synthesis function $h_p(t)$, we can readily compute the adaptive coefficients B_p, via the regular inner prod-uct operation, i.e.,

$$B_p = \int s_p(t) h^*_p(t) dt = \left(\frac{\alpha_p}{\pi}\right)^{1/4} \int s_p(t) \exp\left\{-\frac{\alpha_p}{2}(t-T_p)^2\right\} \exp\{-j\Omega_p t\} dt \qquad (8.25)$$

which guarantees that B_p indeed reflects the signal's local behaviors.

Substituting (8.23) into (8.20) obtains the Gaussian function-based adap-tive spectrogram, such as

$$AS(t, \omega) = 2\sum_p |B_p|^2 \exp\left\{-\left[\alpha_p(t-T_p)^2 + \frac{1}{\alpha_p}(\omega-\Omega_p)^2\right]\right\} \qquad (8.26)$$

Apparently, it is non-negative.

Because the time and frequency resolutions of the Gaussian function are determined by a single parameter, α_p, the computations of the optimal $h_p(t)$ become rather simple. We shall discuss the optimization algorithm shortly.

8.3 Estimation of the Optimal $h_p(t)$ of AGR

The formulae (8.24) and (8.26) reveal that the AGR and adaptive spectrogram can be readily calculated once the elementary functions $\{h_p(t)\}$ are determined. The question then is the selection of the elementary functions $\{h_p(t)\}$. As introduced earlier, $h_p(t)$ should best match $s_p(t)$, in the sense of [Eq. (8.3)]:

$$|B_p|^2 = \max_{h_p} |\langle s_p(t), h_p(t)\rangle|^2$$

Consider that $h_p(t)$ is the Gaussian function defined in (8.22), (8.3) is the optimization with respect to a triplet of $(\alpha_p, T_p, \Omega_p)$. In general, there is no analytical solution of (8.3). Analogous to the zooming process, however, (8.3) can be well approximated numerically.

First, let's digitize (8.3) as

$$|B_p|^2 = \max_{h_p} |\langle s_p(t), h_p(t)\rangle|^2$$

$$= \max_{h_p} \left| \sum_i s_p[i]\left(\frac{\alpha_p}{\pi}\right)^{\frac{1}{4}} \exp\left\{-\frac{\alpha_p}{2}(i - m\Delta M_p)^2\right\} \exp\left\{-j\frac{2\pi n\Delta N_p}{L}i\right\} \right|^2 \tag{8.27}$$

To reduce computation, we limit the time and frequency center in the discrete grids $(m\Delta M_p, n\Delta N_p)$. ΔM_p and ΔN_p denote the time and frequency intervals of the grid. The smaller the ΔM_p and ΔN_p are, the denser the grid. L denotes the effective length of the Gaussian function $h_p[i]$ with the largest variance. Apparently, at each fixed α_p, (8.27) is a regular short-time Fourier transform. When α_p is smaller, (8.27) has a bad time resolution. In this case, $|B_p|^2$ will be less sensitive to the change of $m\Delta M_p$. On the other hand, when α_p is smaller, (8.27) has a good frequency resolution. Consequently, $|B_p|^2$ is very sensitive to the change of $n\Delta N_p$.

Let's define

$$R_p[i] \equiv s_p[i]\left(\frac{\alpha_p}{\pi}\right)^{\frac{1}{4}} \exp\left\{-\frac{\alpha_p}{2}(i - m\Delta M_p)^2\right\} \tag{8.28}$$

Then, (8.27) reduces to the regular L-point DFT, i.e.,

$$|B_p|^2 = \max_{\Delta M_p, \Delta N_p, \alpha_p} \left| \sum_i R_p[i] W_L^{-n\Delta N_p i} \right|^2 \tag{8.29}$$

Now, the matching process can be described as follows. At stage p, first start with a smaller α^0, larger time interval ΔM^0, and smaller frequency interval ΔN^0. Apply FFT to compute (8.29) and find the maximum, say $|B^0|^2$, and corresponding time index $m^0\Delta M^0$. Then, increase the variance parameter to α^1 and frequency interval to ΔN^1 and reduce the time interval ΔM^1 to compute (8.29) in the vicinity of $m^0\Delta M^0$, and find the maximum $|B^1|^2$ and corresponding time

index $m^1 \Delta M^1$. Repeat this process until $|B^i|^2$ no longer increases, for $i = 0,1,2,....$
Then, the optimal variance, time center, and frequency center are

$$\alpha_p \approx \alpha^i \qquad T_p \approx m \Delta M_p = m^i \Delta M^i \qquad \Omega_p \approx 2\pi \frac{n \Delta N_p}{L} = 2\pi \frac{n^i \Delta N^i}{L}$$

and

$$B_p = B^i$$

Fig. 8–7 illustrates the matching process described above, which is very similar
to the zooming operation; first, coarse-search in a larger area. Then, reduce the
searching region and use a smaller scale to zoom in on the object of interest.
Because the basic operations are FFT, the matching process described above is
very efficient. The accuracy of the approximation mainly depends on the size of
time interval ΔM and frequency interval ΔN that are used. The finer the intervals are, the better the accuracy. On the other hand, the finer the intervals are,
the more computations involved. Therefore, there is a trade-off between the
approximation accuracy and computation efficiency.

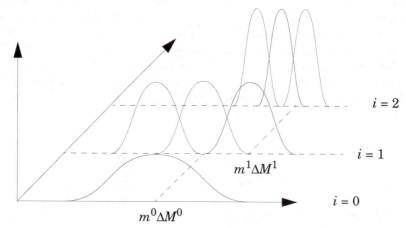

Fig. 8–7 The matching process starts from a coarse variance and time step ΔM^0 (large
ΔM^0) and fine frequency interval ΔN^0. After the match is found at $(m^0 \Delta M^0, n^0 \Delta N^0)$,
the variance and time interval ΔM^1 will be reduced and the frequency interval ΔN^1
will increase. Then, continue the matching process around $m^0 \Delta M^0$ to find the next
matching point $(m^1 \Delta M^1, n^1 \Delta N^1)$. Repeat this process until the projection coefficient B^i
no longer increases.

Once the best match $h_p[i]$ is found, we can compute the residual by $s_{p+1}[i] = s_p[i] - B_p h_p[i]$. Then, we apply the same process to compute the next elementary
function $h_{p+1}[i]$. The process continues until $\|s_{p+1}[i]\| < \varepsilon$, where ε is a predeter-
mined error threshold.

Although $h_p[i]$ computed by the procedure introduced in this section may
not be optimal, we can still use them in the AGR. The convergence of the residual
$\|s_{p+1}[i]\|$, as shown in Section 8.1, is independent of the manner of the selection

of the elementary functions. The optimal $h_p[i]$, however, will yield a fast convergent.

8.4 Comparison of AGR and Other Time-Frequency Representations

To illustrate the usefulness and power of the AGR and adaptive spectrogram, let's look at the signal

$$s(t) = \delta(t - t_1) + \delta(t - t_2) + \exp\{j\omega_1 t\} + \exp\{j\omega_2 t\} \tag{8.30}$$

which is constituted by two time pulses and two frequency pulses. The corresponding STFT spectrogram, scalogram, and adaptive spectrogram are illustrated in Fig. 8–8.

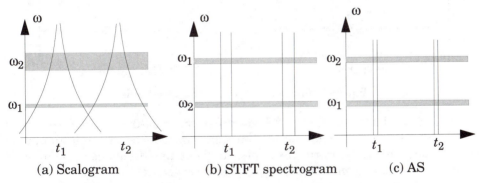

(a) Scalogram   (b) STFT spectrogram   (c) AS

Fig. 8–8 Unlike the wavelet that provides good time resolution at higher frequency and good frequency at lower frequency, the adaptive spectrogram offers good frequency resolution at any frequency band.

Because the wavelet uses wideband windows at the higher frequency band and narrowband windows at the lower frequency band, the wavelet transform offers an excellent time resolution and bad frequency resolution at higher frequency band or vice versa. Therefore, the wavelet is ideal for constant Q analysis such as the octave analysis commonly used in machine vibrational analysis.

The spectrogram by short-time Fourier transform (window FFT) offers a uniform time and frequency resolution at any frequency band. Once the windowing function is selected, so is the frequency resolution. Compared with the wavelet transform, the STFT spectrogram can't offer very high time resolution at a higher frequency band without sacrificing frequency resolution. Similarly, the STFT spectrogram can't offer very good frequency resolution at low frequency without sacrificing time resolution. This is the reason in the first place for researchers to investigate the adaptive spectrogram.

The adaptive spectrogram is rather flexible. By selecting a different variance, one can obtain good frequency resolution at any frequency band. The same is true for the time resolution.

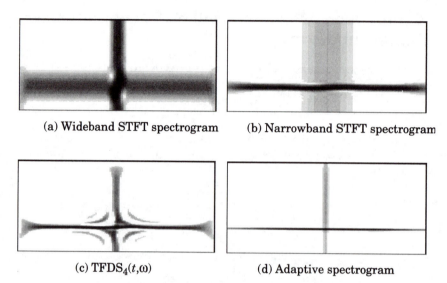

(a) Wideband STFT spectrogram   (b) Narrowband STFT spectrogram

(c) TFDS$_4(t,\omega)$ (d) Adaptive spectrogram

Fig. 8–9 Sinusoidal function plus a time pulse.

Fig. 8–9 illustrates the STFT spectrogram, time-frequency distribution series, and adaptive spectrogram for the signal introduced in the beginning of this chapter. Obviously, in this example, the joint time-frequency resolution of the adaptive spectrogram is superior to all other schemes.

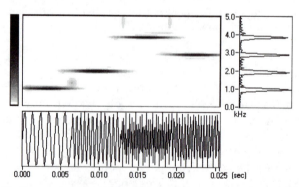

Fig. 8–10 Frequency hopper signals computed by adaptive spectrogram.

Fig. 8–10 and Fig. 8–11 plot the adaptive spectrogram, WVD, STFT spectrogram, and the fourth-order TFDS of a frequency hopper signal. As expected, the WVD suffers from the cross-term interference problem. Both the STFT and the fourth-order TFDS give a good and easy-to-understand spectrogram, but they fail to compare with the adaptive spectrogram for the same signal. The adaptive spectrogram not only shows good time-frequency resolution, but also does not suffer cross-term interference. In addition, it is non-negative.

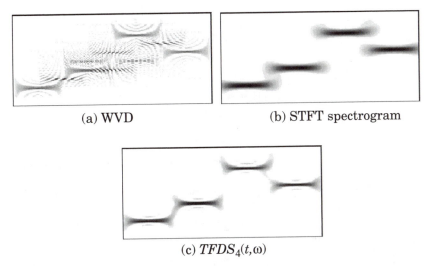

(a) WVD   (b) STFT spectrogram

(c) $TFDS_4(t,\omega)$

Fig. 8–11 Frequency hopper signals.

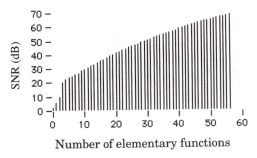

Number of elementary functions

Fig. 8–12 The SNR exponentially increases. With 40 elementary functions, the SNR is improved more than 50 dB.

Fig. 8–12 demonstrates the relationship between the number of elementary functions and the residual error. At pth stage, the SNR is defined by

$$\text{SNR} = 10 \log\left\{\frac{\|s[i]\|^2}{\|s_{p+1}[i]\|^2}\right\}(\text{dB}) \qquad (8.31)$$

Fig. 8–12 indicates that the SNR exponentially increases. Because the frequency hopper signals in this example are dominated by four tones, there is a very noticeable jump after the first four elementary functions. By four elementary functions, the SNR is improved approximately 20 dB.

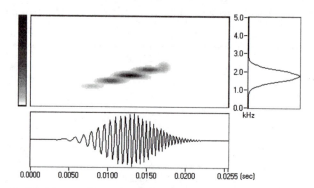

Fig. 8–13 Linear chirp signal computed by adaptive spectrogram.

Fig. 8–13 and Fig. 8–14 plot the adaptive spectrogram, WVD, STFT spectrogram, and the fourth-order TFDS of a linear chirp signal. In this case, the adaptive spectrogram does not offer good time-frequency resolution due to the limited elementary function.

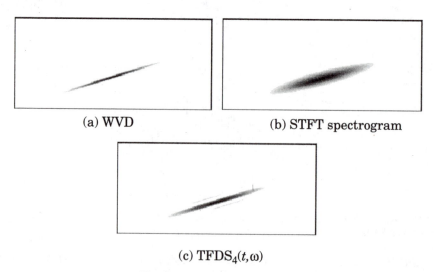

(a) WVD   (b) STFT spectrogram

(c) $\text{TFDS}_4(t,\omega)$

Fig. 8–14 Linear chirp signal.

Fig. 8–10 and Fig. 8–13 are two extreme examples which show the best and the worst scenarios for the adaptive spectrogram. They do point out general guidelines in determining how the adaptive spectrogram will perform. In general, the adaptive spectrogram using the Gaussian functions will perform well if the signal of interest is a combination of short duration pulses with quasi-stationary signals such as the frequency hopper. It does not do well for chirp-type signals.

One way of making the adaptive representation more robust is to use a more general model of the elementary function. For example, use the linear chirp modulated Gaussian function as the basic elementary function, i.e.,

$$h_p(t) = \left(\frac{\alpha_p}{\pi}\right)^{1/4} \exp\left\{-\frac{\alpha_p}{2}(t - T_p)^2\right\} \exp\{j(\Omega_p t + \beta_p t^2)\} \tag{8.32}$$

where the parameter β_p allows us to control the frequency change rate. Fig. 8–15 illustrates the joint time-frequency distribution of $h_p(t)$ in (8.32). It shows that not only can we adjust the variance and time-frequency center, but we can also regulate the orientation of $h_p(t)$ in the joint time-frequency domain by varying the parameter β_p. The linear chirp modulated Gaussian function is more flexible and thereby it could better adapt the analyzed signal. The resulting adaptive spectrogram is

$$AS(t, \omega) = 2\sum_p |B_p|^2 \exp\left\{-\left[\alpha_p(t - T_p)^2 + \frac{1}{\alpha_p}(\omega - \Omega_p - 2\beta_p t)^2\right]\right\} \tag{8.33}$$

which is non-negative.

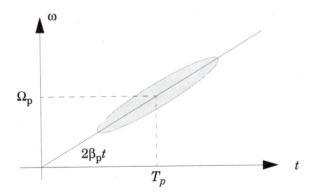

Fig. 8–15 Linear chirp-modulated Gaussian function.

Although the linear chirp modulated Gaussian function possesses some attractive features, the resulting matching process of seeking the optimal $h_p(t)$ is much more complicated. So far, no practical optimization algorithm of computing $(\alpha_p, T_p, \Omega_p, \beta_p)$ has been reported.

Summary

In this chapter, we discussed the adaptive signal expansion and the adaptive spectrogram. The difference between an adaptive signal expansion with other familiar expansion series such as a Fourier expansion is in the selection of the elementary functions. In the adaptive expansion, the elementary functions are highly redundant and they do not usually form a basis. There is quite a degree of freedom to choose the elementary function. Once the models of the elementary functions are selected, the adaptive expansion is done by a repeated matching process. At each step, one elementary function is chosen if it has the maximum similarity with the given signal at that stage. Once the adaptive expansion is achieved, the adaptive spectrogram can be readily computed.

Although the adaptive representation and adaptive spectrogram are independent of the selection of the elementary functions, we focus on the adaptive Gaussian representation. This is because the Gaussian function is optimally concentrated in the joint time-frequency domain. Moreover, the Gaussian-based adaptive spectrogram is non-negative and does not contain the cross-term interference as occurred in many of Cohen's class of distributions.

Compared with the wavelet transforms, the STFT, the adaptive Gaussian representation is more flexible and offers good joint time-frequency resolution. The downside is its computational complexity. Even with the fast algorithm, it is still considerably slower than the STFT. But in applications where the STFT no longer delivers a satisfactory result, the adaptive Gaussian representation is surely a good alternative. In general, the adaptive Gaussian representation and corresponding adaptive spectrogram are especially beneficial to quasi-stationary signals or very short transient signals.

CHAPTER 9

Time-Variant Filter

*I*n addition to the study of a signal's frequency content changes, another very important application of joint time-frequency representations is in the detection and estimation of noise-corrupted signals. While random noise tends to spread evenly into the entire joint time-frequency domain, the signal energy is usually concentrated in a relatively small region. Consequently, the regional signal-to-noise ratio could be substantially improved in the joint time-frequency domain. By applying joint time-frequency representation, we could better extract the noisy signal and reconstruct it. Such processing could be considered as time-variant filtering.

The time-variant filter can be formed based on both linear and bilinear time-frequency representations [161], such as STFT spectrogram [60], scalogram, ambiguity function [173], and Wigner-Ville distribution (see [15], [70], [72], [75], and [109]). Compared to the bilinear presentations, the linear transforms have simple reconstruction structures [43]. Therefore, we focus our discussions on the Gabor expansion-based time-variant filter.

One of the main difficulties in designing the time-variant filter is that modified time-frequency representations usually are not the valid time-frequency representations. In other words, for a given modified time-frequency representation, there may be no physically existing signal that corresponds to it. To overcome this problem, we first introduce the least square error (LSE) method in Section 9.1. According to this method, the time-frequency representation of the estimated signal is closest to, in the sense of the minimum square error, the noise-reduced time-frequency representation. The LSE is one of the most common technologies. It has been extensively studied for many years. But it may not be the best solution for many applications. In particular, to solve the LSE, we need to compute the pseudoinverse, which is demanding in terms of computation and memory. In many real applications, the data size often is larger than 10,000. In those cases, it is difficult to apply an LSE algorithm if only conventional personal computers are available.

As an alternative, in Section 9.2, we present an iterative algorithm. As numerical simulations indicated, the iterative algorithm not only yields a better signal-to-noise ratio, but also is amendable for real-time implementation.

9.1   LSE Filter

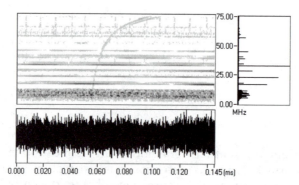

Fig. 9–1 Due to the low SNR, the ionized impulse signal cannot be recognized in either time or frequency domains. However, by joint time-frequency representations, we can readily distinguish it. (Data courtesy of the Non-Proliferation & International Security Division, Los Alamos National Laboratory.)

In addition to the area of signal analysis, joint time-frequency representations are powerful for the detection and estimation of the noise-corrupted signal. Fig. 9–1 depicts the impulse signal received by the U.S. Department of Energy ALEXIS/BLACKBREAD satellite. After passing through dispersive media, such as the ionosphere, the impulse signal becomes a non-linear chirp signal. As shown in Fig. 9–1, while the time waveform is severely corrupted by random noise, the power spectrum is mainly dominated by radio carrier signals that basically are unchanged over time. In this case, neither time waveform nor the power spectrum indicate the existence of the impulse signal. However, when looking at the joint time-frequency plot, we could immediately identify the presence of the chirp-type signal arching across the joint time-frequency domain.

In general, random noise tends to spread evenly in the joint time-frequency domain, while the signal itself concentrates in a relatively small range. Consequently, by joint time-frequency representation, the signal-to-noise ratio (SNR) could be substantially improved. Once the interesting signal has been identified, we can mask it from the background noise and reconstruct it.

The above statement applies for any joint time-frequency representations; both linear and bilinear. For bilinear representations, there is no phase information. The reconstruction is always troublesome. On the other hand, the linear representations have a simple inverse structure. Therefore, we focus our discussion on the Gabor expansion-based time-variant filter.

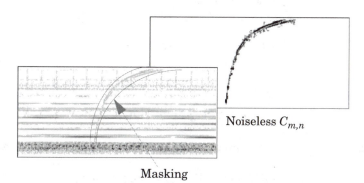

Masking

Fig. 9–2 Masking the desired signal from joint time-frequency representation.

As shown in Fig. 9–2, the typical procedure of the time-variant filter is first to take the Gabor transform $G\vec{x}$, where $\vec{x}$ denotes the noise-corrupted signal vector. Then, mask the desired signal portion from the background noise to obtain noiseless Gabor coefficients

$$\hat{c} = \Phi G\vec{x}, \tag{9.1}$$

where Φ denotes a mask function. Finally, apply the Gabor expansion to compute the noise-reduced time waveform

$$\vec{s} = H\Phi G\vec{x}. \tag{9.2}$$

As discussed previously, the modified Gabor coefficients in general are not valid Gabor coefficients. In other words, the modified Gabor coefficients may not correspond to any physically existing time functions, i.e.,

$$G\vec{s} \neq \hat{c} \tag{9.3}$$

In this case, we have to estimate the time waveform that is the best in some sense. The most common criterion is the least square error (LSE) method, which finds the time waveform whose Gabor coefficients are closest to the desired Gabor coefficients, in the sense of

$$\xi = \min_{\hat{s}} \left\| \hat{c} - G\hat{s} \right\|^2 \tag{9.4}$$

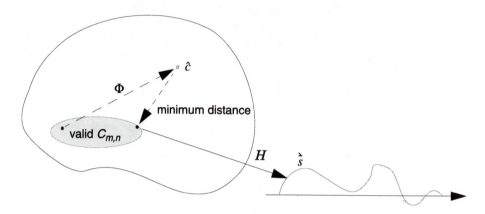

Fig. 9–3 Map of LSE filtering.

For the discrete-time samples, G is a matrix whose entries are given by

$$\gamma_{mN+n,k} \equiv \gamma^*[k-m\Delta M]W_N^{-nk} \qquad \text{for } 0 \le n < N \tag{9.5}$$

The vector $\hat{c}$ is formed by masked Gabor coefficients, i.e.,

$$c_{mN+n} = \hat{C}_{m,n} = \left\{ \begin{array}{ll} C_{m,n} & \phi_{m,n}=1 \\ 0 & \phi_{m,n}=0 \end{array} \right. \tag{9.6}$$

where $\phi_{m,n}$ denote the entries in the binary mask matrix Φ. The physical interpretation of (9.4) is illustrated in Fig. 9–3. The Gabor coefficients of the solution $\hat{s}$ of (9.4) are most similar to the desired Gabor coefficients $\hat{c}$.

From the elementary matrix analysis, the solution of (9.4) is the pseudoinverse of G, that is,

$$\hat{s} = (G^T G)^{-1} G^T \hat{c} = A^{-1}\hat{d} \tag{9.7}$$

where

$$A = G^T G \qquad \hat{d} = G^T \hat{c}$$

The entries of the matrix A are

$$a_{k,k'} = \sum_m \sum_n \gamma[k-m\Delta M]\gamma^*[k'-m\Delta M]W_N^{-n(k-k')}$$

$$= \frac{1}{N}\delta(k-k'-qN)\sum_m \gamma[k-m\Delta M]\gamma^*[k'-m\Delta M]$$

If the length of $\gamma[k]$ is L, then,

$$a_{k,k'} = \begin{cases} \dfrac{1}{N}\sum_m \gamma[k - m\Delta M]\gamma^*[k' - m\Delta M] & |k - k'| = qN < L \\ 0 & \text{otherwise} \end{cases} \tag{9.8}$$

Note that $a_{k,k'} = a^*_{k',k}$. If $\gamma[k]$ is real, then A is real, symmetrical, and a sparse matrix.

The elements of the vector $\vec{d}$ in (9.7) are

$$d_k = \sum_m \sum_n \hat{C}_{m,n}\gamma[k - m\Delta M]W_N^{nk} = \sum_m \gamma[k - m\Delta M]\sum_{n=0}^{N-1}\hat{C}_{m,n}W_N^{nk} \tag{9.9}$$

In most applications, the signal is real-valued, then

$$\hat{C}_{m,n} = \hat{C}^*_{m,N-n} \qquad \text{for } 0 \le n < \frac{N}{2} \tag{9.10}$$

Consequently, the vector $\vec{d}$ is also real-valued. In this case, (9.7) can be solved by the real matrix operation.

When $\hat{C}_{m,n}$ are the valid Gabor coefficients for the sequence $f[k]$, that is,

$$\hat{c} = G\vec{f} \tag{9.11}$$

substituting (9.11) into (9.7) yields

$$(G^TG)^{-1}G^T\hat{c} = (G^TG)^{-1}G^TG\vec{f} = \vec{f} \tag{9.12}$$

In this case, the square error $\xi = 0$ in (9.4).

At first glance, it seems that we can use the arbitrary analysis function $\gamma[k]$ and sampling steps, ΔM and ΔN, to implement the time-variant filter. It is not true in fact. To ensure the existence of the inverse of A in (9.7), $\gamma[k]$ and sampling steps have to meet certain conditions. An imprudent choice of $\gamma[k]$ and sampling steps may result in A being non-invertible and numerically unstable.

The key step of the LSE filtering introduced above is to solve the pseudoinverse of matrix A in (9.7). Matrix A is a square matrix whose number of rows/columns at least is equal to the number of samples (depending on the oversampling rate). When data are large, it would be difficult to solve A^{-1} by conventional digital computers. For example, the number of data in Fig. 9–1 is more than 9,000. For the double oversampling, the size of matrix A is approximately 20,000 - by - 20,000. Unless using supercomputers, it is impossible to compute the pseudoinverse of such a large matrix.

Although the LSE is the most well-known error measure, it may not be the one that always fits the problem at hand. In many real applications, we often are more interested in the SNR. As an alternative, we shall introduce an iterative time-variant filter in the next section.

9.2   Iterative Time-Variant Filter

The iteration approach can be described as follows. First, map the noisy signal into the joint time-frequency domain via the Gabor transform G to obtain the Gabor coefficient vector $G\vec{x}$. If we do not alter the Gabor coefficients, then we should be able to recover the original signal by the Gabor expansion H, that is,

$$HG\vec{x} = \vec{x} \tag{9.13}$$

If we apply a mask function Φ to filter out some noise and compute the Gabor expansion, then

$$\vec{x_1} = H\Phi G\vec{x} \tag{9.14}$$

Because the modified Gabor coefficients $\Phi G\vec{x}$ generally are no longer the valid Gabor coefficients, that is,

$$G\vec{x_1} \neq \Phi G\vec{x} \tag{9.15}$$

$\vec{x_1}$ may not possess the desired time-frequency properties.

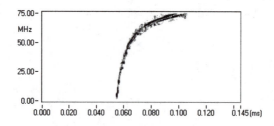

Fig. 9–4 Masked Gabor coefficients of ionized impulse signal in Fig. 9–1.

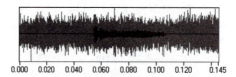

Fig. 9–5 Comparison of estimation and noise corrupted signals.

Now, let's repeat the process described by (9.14) and (9.15), and continue this process; after ith iterations, we have

$$(H\Phi G)^i\vec{x} = \vec{x_i} = \hat{s} \tag{9.16}$$

From the matrix analysis theory, we know that the iteration in (9.16) converges as long as the maximum eigenvalue of the matrix $H\Phi G$ is less than one. We can further prove [194] that, under certain conditions such as when the dual functions $h[k]$ and $\gamma[k]$ are identical, the first iteration $\hat{x}_1$ is equivalent to the LES solution. As the number of iterations increases, the time waveform gets smooth and $G\hat{x}_i$ will converge to inside of the masked area.

Fig. 9–4 plots the masked Gabor coefficients for the example illustrated in Fig. 9–1. Fig. 9–5 depicts the reconstructed signal after five iterations. In this example, the number of samples is more than 9,000, which is impossible to be solved by the LES method introduced in Section 9.1 due to the computational complexity and memory limitations.

It is well known that the continuous-time signals cannot be finitely supported in time and frequency simultaneously. For the discrete-time sequence or the continuous-time signal that is measured in the discrete time-frequency grids, they could have finite non-zero transform coefficients.

For example, $h[k]$ and $\gamma[k]$, illustrated in Fig. 9–6, are biorthogonal to each other, i.e.,

$$\langle \tilde{h}_{m,n}[k], \tilde{\gamma}[k] \rangle = \delta[m]\delta[n] \tag{9.17}$$

Assume that the signal $s[k] = h[k]$, then the corresponding Gabor coefficients are

$$C_{m,n} = \langle \tilde{s}[k], \tilde{\gamma}_{m,n}[k] \rangle = \langle \tilde{h}[k], \tilde{\gamma}_{m,n}[k] \rangle = \delta[m]\delta[n] \tag{9.18}$$

which indicates that the Gabor coefficients are a pulse at $(0,0)$, the origin of the joint time-frequency domain.

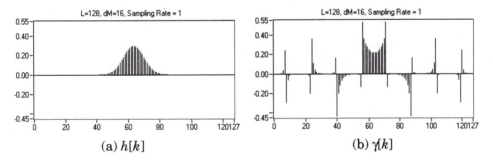

(a) $h[k]$ (b) $\gamma[k]$

Fig. 9–6 $h[k]$ and $\gamma[k]$ are biorthogonal.

Another trivial example is that the analysis function $\gamma(t)$ is a rectangular pulse, such as

$$\gamma(t) = \begin{cases} 1 & 0 \le t < T \\ 0 & \text{otherwise} \end{cases} \tag{9.19}$$

Assume that signal $s(t) = \gamma(t)$, then the Gabor coefficients are

$$C_{m,n} = \langle s, \gamma_{m,n} \rangle = \langle \gamma, \gamma_{m,n} \rangle$$

$$= \int \gamma(t)\gamma^*(t - mT)\exp\{-jn\Omega t\}dt$$

$$= \delta(m)\delta(n)$$

which indicates that, except for $C_{0,0}$, all other Gabor coefficients are equal to zero.

For the LSE method, we chose a solution $\hat{s}$ such that its Gabor coefficients are optimally close to the modified Gabor coefficients $\Phi G\hat{x}$. Although the Gabor coefficients of the estimated signal have minimum distance to the desired Gabor coefficients, in general they do not completely fall into the desired time-frequency region Φ. On the other hand, the solution given by the iteration method ensures that the time-frequency support of $\hat{s}$ indeed is inside of the desired region. Although in this case, the error Γ, where

$$\Gamma = \left\| G\hat{s} - \Phi G\hat{x} \right\|^2 = \left\| G(H\Phi G)^i\hat{x} - \Phi G\hat{x} \right\|^2 \tag{9.20}$$

is not the minimum for $i > 1$, numerical simulations indicate that the SNR is improved as the number of iterations increases. The simulations show that the peak SNR always occurs for i around 5.

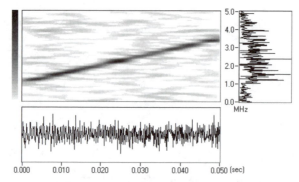

Fig. 9–7 Noise-corrupted linear chirp signal (SNR = 0 dB).

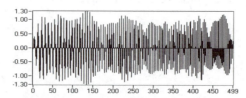

Fig. 9–8 The improvement is more than 15 dB.

Fig. 9–7 plots a noise-corrupted linear chirp signal, in which the SNR is close to 0 dB. Due to the low SNR, it is difficult to extract the signal from either the time or frequency domain. To effectively remove the noise, we apply the iterative time-variant filter. Fig. 9–8 plots the reconstructed signal. In this case, the improvement of SNR is more than 15 dB.

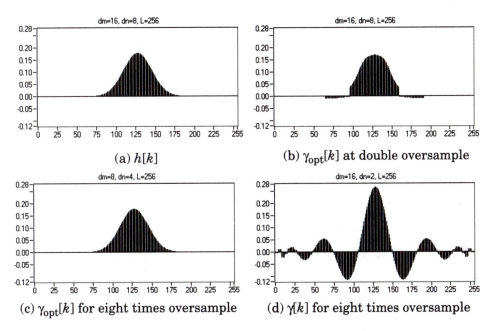

(a) $h[k]$ (b) $\gamma_{opt}[k]$ at double oversample

(c) $\gamma_{opt}[k]$ for eight times oversample   (d) $\gamma[k]$ for eight times oversample

Fig. 9–9 Dual functions for double oversampling.

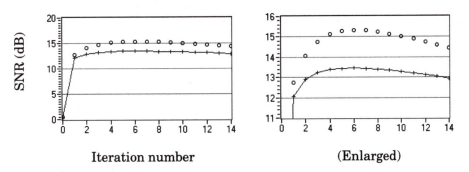

Iteration number   (Enlarged)

Fig. 9–10 SNR vs. iteration number. (The right plot is the enlarged part of the left plot; "+" = double oversampling; "o" = eight times oversampling.)

Fig. 9–9 plots the synthesis function $h[k]$ and different analysis functions $\gamma[k]$ used in this example. The comparison is illustrated in Fig. 9–10, where "+" and "o" represent the results obtained by dual functions $\gamma_{opt}[k]$ in Fig. 9–9b and c, respectively. Because $\gamma_{opt}[k]$ in Fig. 9–9c (corresponding to eight times oversampling) is closer to $h[k]$ in Fig. 9–9a than $\gamma_{opt}[k]$ in Fig. 9–9b (corresponding to double oversampling), the performance of "o" is better than "+." As mentioned earlier, the results obtained by the first iterations are equivalent to LSE. When the number of iterations increases, SNR is substantially improved. For the case of $\gamma_{opt}[k]$ in Fig. 9–9c, the peak SNR is 15.5 dB, that is, approximately 25% higher than that achieved by LES.

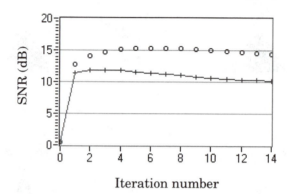

Iteration number

Fig. 9–11 "+" corresponds to $\gamma[k]$ in Fig. 9–9d; "o" corresponds to $\gamma_{opt}[k]$ in Fig. 9–9c. (Although both $\gamma[k]$ are for eight times oversampling, the orthogonal-like dual function $\gamma_{opt}[k]$ yields much better results.)

Fig. 9–11 illustrates the impact of oversampling. Although both dual functions used work at eight times oversampling, the orthogonal-like dual function $\gamma_{opt}[k]$ yields much better SNR. This example shows that the outcome of the iterative time-variant filter is related to the closeness between analysis and synthesis functions rather than the oversampling rate.

For computation considerations, we would like the sampling rate as low as possible. On the other hand, to ensure that $h[k]$ and $\gamma[k]$ are similar requires oversampling. There is a trade-off between the computational burden and the performance of the time-variant filter. For the given oversampling rate, the orthogonal-like Gabor expansion introduced in Chapter 3 produces the $\gamma_{opt}[k]$ that is most similar to $h[k]$.

Summary

In this chapter, we briefly introduced the concept and implementation of the time-variant filter. The time-variant filter has been found to be extremely powerful for wideband and non-stationary signal estimation. First, we introduced the LSE algorithm. Although the LSE method is the most well-known approach, it may not be suitable for some real applications in which the size of samples is huge. As an alternative, we introduced the iterative method. The iterative time-variant filter not only is computationally efficient, but also offers better SNR. Our experiences indicated, in terms of SNR, that the performance of the iterative time-variant filter sometimes could be 20 ~ 30% better than that accomplished by the LSE method.

Applications of JTFA

To demonstrate the effectiveness of JTFA, in what follows we shall introduce some real application examples in the areas of radar, medical, and economic data analysis. Although the applications discussed in this part seem rather specific, the ideas and methodologies behind these applications in fact are general for all those who are interested in using JTFA to solve their own problems.

In Chapter 10, we discuss the application of JTFA for radar image processing. The radar may be one of the oldest areas using JTFA. Many JTFA techniques were initially motivated by radar applications. Radar systems discussed in this book, however, are much more advanced than those used in World War II. Unlike those early radar systems, not only are advanced radars able to measure the distance from the target, but they also provide the target image. These advanced imaging radars have been widely used for aerial mapping and target identification under all weather conditions.

Imaging radar is also called Doppler-range radar, which is made up of Doppler spectra in each range cell. A main issue in the radar community has been image quality. Conventional radar processing uses DFT to compute Doppler frequencies. Due to a target's irregular motion, the Doppler frequency is of time-varying nature. Consequently, Fourier transform-based radar systems often yield blurred images.

Since the beginning of developing imaging radar, many algorithms have been proposed for focusing and compensating a target's complex motion to resolve the image blurring problem. Most motion-compensation algorithms, however, are very complicated. Recently, Chen [22] introduced an approach using joint time-frequency transform, introduced in Chapter 7, to lift the restrictions of the Fourier transform. Replacing the conventional Fourier transform with a joint time-frequency transform, a high-resolution radar image can be achieved without applying complicated motion-compensation algorithms. The joint time-frequency processing-based radar imaging is especially useful for resolving the image blurring in the cross-range.

Another type of smearing in the radar image is caused by so-called non-point scattering. When cavities or duct-type structures are present in targets, radar images can be plagued by artifacts due to the complex scattering behaviors in these structures. These scattering mechanisms appear in the radar image as blurred clouds which extend down-range and do not correspond to the spatial features on the target. On the other hand, the same mechanisms do provide important features which, if properly interpreted, can become important factors in the target classification process. Trintinalia and Ling [178] first successfully applied the adaptive Gaussian basis representation, introduced in Chapter 8, to decompose the down-range samples of the complex radar image. The portion of the signal represented by wideband basis functions is used to reconstruct the radar image, and the portion corresponding to narrowband basis functions is utilized for resonant feature identification. The decomposition scheme introduced by Trintinania and Ling not only substantially improves the radar image by removing the blurring clouds, but also reinterprets the information contained in these clouds in an alternative feature space to facilitate target recognition.

In Chapter 11, we introduce JTFA for localization of brain functions. The human brain is often considered to be the most complex biological structure in existence. It consists of large numbers of neurons and interconnections which process and transmit information. Understanding the working brain and nervous system has been an active subject of research across a wide variety of disciplines. The result presented in Chapter 11 is related to a problem of major significance in both basic and clinical neuroscience - the determination of the location of functional areas related to the generation of electrical activity in the human brain.

One major challenge of localizing brain functions lies in the low signal-to-noise ratio of collected brain signal. The electrical brain signal usually is embedded in background noise. Because of its time-varying nature, it is very difficult to effectively extract useful electrical brain signals from background noise with conventional filters. Applying the time-varying filter, introduced in Chapter 9, Sun and Sclabassi significantly improved the accuracy of localizations. The solution of this problem will provide the capability to answer important questions, by a non-invasive means, concerning activated areas of the brain where information is processed. In addition, the identification of active neural substrates will aid in making critical neurosurgical decisions.

In Chapter 12, we briefly discuss a most interesting and controversial problem, economic data analysis. Traditionally, the economic system is studied by probability theory. Economists generally believe that economic data are pure random walk. Economic movements are toward equilibrium. The stock price disequilibrium cannot last long in an efficient market. A reasonable goal of buy-and-hold investment is the average return.

The recent development of nonlinear dynamics suggests that business cycles may be better understood and modeled by the concept of color chaos. Applying JTFA, Chen demonstrates the existence of stable business cycles. This observation not only helps us to better understand the economic system, but also is significant in utilizing chaos theory to study economic problems. Although JTFA remains a buzzword to most economists, it has been found to have great potential in the areas of finance and economics.

JTFA provides a powerful tool for many applications. To solve a real-world problem, however, it is necessary to fully understand its mechanism and concepts. For example, what can JTFA offer? Why does the conventional technique not work? All contributors in this part are not only experts in their own areas, but also understand JTFA well. We have tried to write this part in a style that appeals to general readers rather than experts in those particular fields. We are careful to explain the motivation for using JTFA, while avoiding lengthy background discussions. Hopefully, readers will find this part both enjoyable and enlightening.

CHAPTER 10

Applications of JTFA to Radar Image Processing

*R*adar has long been used for the detection and measuring of targets. The advances in radar signal processing further permit radar to make images of terrain as well as moving targets. Imaging radar has been successfully used to produce high-resolution imagery for aerial mapping and target identification in all weather conditions.

Radar transmits electromagnetic waves and processes returned waves from a target to produce a high-resolution image of the target. A target usually consists of a number of scatterers. They can be discontinuities, corners, or cavities in the target. Each type of scatterer has a different backscattering behavior. This provides a way to identify the target based on its backscattering behavior.

To obtain a high-resolution radar image, a wide signal bandwidth and longer imaging time are required. However, due to the time-varying behavior of the returned signal and due to multiple scattering behaviors of the target, the radar resolution can be significantly degraded and the image becomes blurred.

Radar utilizes Doppler information in obtaining the cross-range resolution. The conventional radar processing uses the Fourier transform to obtain Doppler frequencies. In order to apply the Fourier transform properly, the Doppler frequency contents of the radar data should be stationary during the imaging time interval. Otherwise, the Fourier transform can cause the Doppler frequency spectrum to become smeared. Since the beginning of the development of the imaging radar, many algorithms have been proposed for focusing and compensating a target's complex motion to resolve the image blurring problem. Most algorithms are based on the use of the Fourier transform for image construction. However, for utilizing the Fourier transform adequately, some restrictions must be applied. To satisfy these restrictions and to obtain high-resolution images, the motion-compensation algorithm can be very complicated. Recently, Chen [22] introduced an approach using joint time-frequency transform to lift the restrictions of the Fourier transform. Replacing the conventional Fourier transform with a joint time-frequency transform, a high-resolution radar image can be achieved without applying complicated motion-compensation algorithms. The joint time-frequency processing-based radar imaging is especially useful for resolving the image blurring in the cross-range.

When cavities or duct-type structures are present in targets, radar images can be plagued by artifacts, due to the complex scattering behaviors in these structures. These scattering mechanisms appear in the radar image as blurred clouds which extend down-range and do not correspond to the spatial features on the target. On the other hand, the same mechanisms do provide important features which, if properly interpreted, can become important factors in the target classification process. Trintinalia and Ling [178] first successfully applied the adaptive Gaussian basis representation to decompose the down-range samples of the complex radar image. The portion of the signal represented by wideband basis functions is used to reconstruct the radar image, and the portion corresponding to narrowband basis functions is utilized for resonant feature identification. The decomposition scheme introduced by Trintinalia and Ling not only substantially improves the radar image by removing the blurring clouds, but also reinterprets the information contained in these clouds in an alternative feature space to facilitate target recognition.

The first section of this chapter mainly deals with the blurred image caused by irregular translational and rotational motions. Section 10.2 discusses the extraction of backscattering features from complex scattering mechanisms.

10.1   Radar Range-Doppler Imaging[*]

Radar transmits electromagnetic waves to an object which consists of a number of point scatterers and receives the scattered waves from the object. In the radar receiver, the returned signal from the object is the sum of the returned signals from the scatterers of the object. The scattering properties of the object describe the features of the object. Usually, objects have scatterers with a variety of backscattering behaviors. Scatterers can be discontinuities, corners, or cavities. The backscattering from the scattering centers of the object can be simple specular or diffractive scattering from edges or multiple specular scattering from cavities. Since the integrated effect of the scattered fields can be measured directly by the radar, the spatial distribution of the reflectivity corresponding to the object can be constructed by a radar processor. The distribution of the reflectivity is referred to as the radar image of the object. The object's reflectivity is usually

---

[*] Contributed by Victor C. Chen, Naval Research Laboratory, Washington D.C. 20375.

mapped onto a range (sometimes called the *slant-range* or *down-range*) and the cross-range plane and is viewed as a radar image of the object. The range is the dimension along the radar's line-of-sight to the object. The cross-range is the dimension transverse to the line-of-sight.

Because the radar image presents a spatial distribution of the object's reflectivity, it should not be judged solely by its similarity to the visual image of the same object. However, a useful radar image must represent the spatial distribution of the radar's reflectivity faithfully. Therefore, high-resolution radar images are always demanded. The range resolution is directly related to the bandwidth of the transmitted radar signal. Stepped-frequency waveforms and frequency-modulated chirp waveforms are examples of wideband radar signals, which are commonly used in radar imaging systems to achieve high range resolution. The cross-range resolution is determined by the antenna beamwidth, which is inversely proportional to the length of the antenna aperture. Thus, a larger antenna aperture can provide higher cross-range resolution. To achieve high cross-range resolution without using a large antenna aperture, synthetic array processing is widely employed. Synthetic array radar processing coherently combines signals obtained from sequences of small apertures to emulate the results from a large aperture.

Synthetic array radar includes both synthetic aperture radar (SAR) and inverse synthetic aperture radar (ISAR). Traditionally, SAR refers to the situation in which the radar is moving and the object is stationary; ISAR refers to the geometrical inverse in which the object is moving and the radar is stationary. For ISAR, the synthetic aperture is formed by coherently combining signals obtained from a single aperture as it observes a rotating object. The rotation of the object emulates the result from a larger circular aperture focusing at the rotation center of the object.

The idea of ISAR imaging is to use Doppler information to obtain the cross-range resolution. Due to the object's rotation, which can be characterized as a superposition of pitch, roll, and yaw motions, different parts of the object have slightly different velocities relative to the radar and, hence, produce slightly different Doppler frequencies in the radar receiver. The differential Doppler shift of adjacent point scatterers can be observed in the receiver; therefore, the distribution of the object's reflectivity can be measured by the Doppler spectrum. The conventional method to retrieve Doppler information is the Fourier transform. In order to use the Fourier transform properly, some restrictions must be applied. The Doppler frequency contents of the data should not change within the time duration of the data. If the Doppler contents do change with time, the Doppler spectrum obtained from the Fourier transform becomes smeared, and, thus, the cross-range resolution is degraded. However, the restrictions can be lifted if the Doppler information can be retrieved with a method other than the Fourier transform that does not require stationary Doppler frequencies.

The purpose of this section is to discuss how joint time-frequency analysis can be applied to radar imaging to lift the restrictions required by the Fourier transform and to retrieve Doppler information without smearing the image in the cross-range. However, in some cases, the radar images can still be blurred in the down-range direction due to the presence of cavities or other duct-type structures on the target, such as the engine intake of an aircraft. In the next section, we will discuss how joint time-frequency analysis can be applied to extract the blurring clouds due to these complex scattering behaviors and to obtain a clean radar image containing only physically meaningful scattering centers.

Since radar images convey information which may not be obtainable by other imaging means, they are widely used in many areas, such as remote sensing and wide-area surveillance (see [124], [165], and [184]). SAR has been successfully used for ground mapping, and ISAR has shown promising results for imaging and classifying moving targets in all weather conditions.

10.1.1   Synthetic Aperture Radar Imaging of Moving Objects

It is necessary to describe why a radar can generate a range-Doppler image of a moving object before describing how the range-Doppler image is constructed in a radar system.

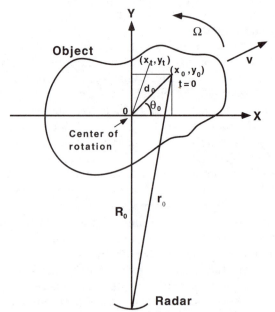

Fig. 10–1 Geometry of the radar image of an object.

The geometry of the radar imaging of an object is shown in Fig. 10–1. Assume at a time $t = 0$ that the distance from the radar antenna to the geometric center of the object is R_0. The object is described in Cartesian coordinates with its origin located at the geometric center. For simplicity, it is only necessary to show a planar object with the radar located on the plane of the object along the y-axis.

First, consider a point scatterer located at (x_0, y_0) at time $t = 0$. The range of the point scatterer from the antenna is

$$r_0 = (R_0^2 + d_0^2 + 2d_0 R_0 \sin\theta_0)^{1/2}$$

where $d_0 = (x_0^2 + y_0^2)^{1/2}$ is the distance of the scatterer from the origin of the object, and $\theta_0 = \arctan(y_0/x_0)$ is the initial rotation angle of the point scatterer.

Assume the radar transmits a sinusoidal waveform with a carrier frequency f:

$$s_t(t) = \exp\{j2\pi ft\}$$

The returned signal from the point scatterers is heterodyned in the receiver. Then, the baseband signal becomes

$$s_r(f, r_0) = \rho(x_0, y_0)\exp\left\{j2\pi f\frac{2r_0}{c}\right\} = \rho(x_0, y_0)\exp\{j\Phi_r(r_0)\}$$

where $\rho(x_0, y_0)$ is the reflectivity of the point scatterer, and the phase of the baseband signal is

$$\Phi_r(r_0) = 2\pi f\frac{2r_0}{c} \tag{10.1}$$

where c is the velocity of the wave propagation.

If the object has only translational motion, the range of the point scatterer at time t becomes

$$r(t) = r_0 + \upsilon_R t$$

where υ_R is the radial velocity of the point scatterer.

When the object is rotating with an angular rotation rate Ω about its geometric center, at time t, the point scatterer is rotated to (x_t, y_t). Then, the range of the point scatterer becomes

$$r_0 = (R_0^2 + d_0^2 + 2d_0 R_0 \sin\theta_t)^{1/2}$$

where $\theta_t = \theta_0 + \Omega t$. When the range of the object is much greater than the radius of rotation, i.e., $R_0 \gg d_0$, the range of the point scatterer becomes

$$r(t) = R_0 + x_0 \sin\Omega t + y_0 \cos\Omega t$$

where

$$x_0 = d_0 \cos\theta_0$$

and

$$y_0 = d_0 \sin\theta_0$$

Thus, when the object has a complex translational and rotational motion, the range of the point scatterer becomes

$$r(t) = r_0 + v_R t + x_0 \sin\Omega t + y_0 \cos\Omega t$$

and the phase of the received signal is

$$\Phi_r(r_t) = 2\pi f \frac{2r(t)}{c} = 2\pi f \frac{2(r_0 + v_R t + y_0 + x_0 \sin\Omega t + y_0 \cos\Omega t)}{c}$$

Because the time-derivative of the phase is frequency, by taking the time-derivative of the phase, the Doppler frequency shift induced by the object's motion is

$$f_D = \frac{1}{2\pi}\frac{d}{dt}\Phi_r(r_t) = \frac{2f}{c}\frac{d}{dt}r(t) = \frac{2f}{c}v_R + \frac{2f}{c}(x_0\Omega - y_0\Omega^2 t) = f_{D_{\text{trans}}} + f_{D_{\text{rot}}}$$

where the Doppler frequency shift induced by the translational motion is

$$f_{D_{\text{trans}}} = \frac{2f}{c}v_R$$

and that induced by the rotational motion is

$$f_{D_{\text{rot}}} = \frac{2f}{c}(x_0\Omega - y_0\Omega^2 t) \tag{10.2}$$

where it has been assumed that Ωt is very small, hence, $\cos\Omega t \cong 1$ and $\sin\Omega t \cong \Omega t$. The first and second terms of (10.2) come from the linear and quadratic parts of the phase function, respectively. Clearly, the quadratic part of the rotational Doppler frequency shift is a function of time. Therefore, the rotational Doppler frequency shift is time-varying, even if the rotation rate is constant. Because the Doppler frequency shift is related to the geometric location of the scatterer (x_0, y_0), another scatterer at a different geometric location in the object may have a different Doppler frequency shift.

Based on the returned signal from a single point scatterer, the returned signal from the object can be represented as the integration of the returned signals from all scatterers in the object:

$$S(f) = \iint \rho(x, y) \exp\left\{-j2\pi f \frac{2r}{c}\right\} dx dy \tag{10.3}$$

The reflectivity function of the object constructed by measuring the Doppler spectrum at each range cell is mapped onto a range-Doppler plane and the radar image becomes a range-Doppler image. Since the Doppler frequency shift can be positive or negative, sometimes the radar range-Doppler image of an object can be upside-down. However, the geometric correspondence between the scatterers within the object and the points on the range-Doppler plane can be found; this correspondence provides a basis for object identification from radar range-Doppler images.

Fig. 10–2 illustrates an ISAR geometry [184] in which an object has translational and rotational motion with a rotation angle in the global (u,v) coordinate system. The range $R(t)$ and the rotation angle $\theta(t)$ of the object are functions of time. They can be determined by initial range R_0 and rotation angle θ_0, initial velocity v_R and angular velocity Ω, initial radial acceleration a_R and angular acceleration α, and other higher order terms:

$$R(t) = R_0 + v_R t + \frac{1}{2} a_R t^2 + \ldots$$

and

$$\theta(t) = \theta_0 + \Omega t + \frac{1}{2} \alpha t^2 + \ldots$$

The range of a point scatterer at (x,y) in the local coordinate system becomes $r(t) = R(t) + x \cos \theta(t) - y \sin \theta(t)$. Substituting this range into (10.3) gives the baseband signal in the receiver as

$$S(f, t) = \exp\left\{-j4\pi f \frac{R(t)}{c}\right\} \iint \rho(x, y) \exp\left\{-j2\pi[x f_x(t) - y f_y(t)]\right\} dx \, dy \qquad (10.4)$$

where the components of the spatial frequency are determined by

$$f_x = \frac{2f}{c} \cos \theta(t) \qquad (10.5)$$

and

$$f_y = \frac{2f}{c} \sin \theta(t) \qquad (10.6)$$

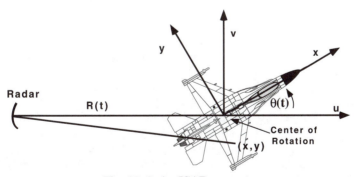

Fig. 10–2 An ISAR geometry.

The objective of radar processing is to estimate the object's reflective density function $\rho(x,y)$ from received baseband signal samples, the so-called frequency signature $S(f,t)$. From (10.4), it is apparent that if the object's range is known exactly and the velocity and acceleration of the object's motion are constant and known exactly over the imaging time duration, then the extraneous phase term of the motion $\exp\{-j4\pi fR(t)/c\}$ can be removed exactly by multiplying $\exp\{j4\pi fR(t)/c\}$ on both sides of (10.4). Therefore, the reflective density function $\rho(x,y)$ of the object can be obtained simply by taking the inverse Fourier transform of the phase-compensated frequency signature $S(f,t)\exp\{j4\pi fR(t)/c\}$. The process of estimating the object's range $R(t)$ and removing the extraneous phase term $\exp\{-j4\pi fR(t)/c\}$ is called focusing, or gross translational motion compensation. Then, the inverse Fourier transform can be used to construct the reflective density function of the object. For SAR, the motion compensation is facilitated by measuring the actual motion of the radar platform. In ISAR, the actual motion can be measured by a range-tracker or estimated by a motion-compensation algorithm which estimates motion parameters and compensates motion with respect to object's range, velocity, acceleration, and other higher order terms.

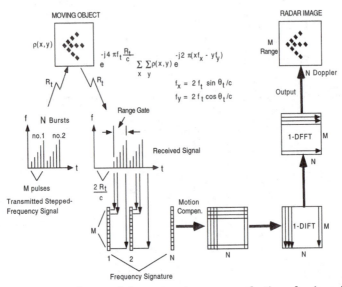

Fig. 10–3 Illustration of stepped-frequency inverse synthetic radar imaging of a moving object.

Now, it can be seen how the radar imaging system is built and how a range-Doppler image is constructed. Fig. 10–3 shows a block diagram of a radar imaging system using stepped-frequency waveforms to illustrate the process of synthetic radar imaging. The stepped-frequency radar transmits a sequence of N bursts. Each burst consists of M narrowband pulses. Within each burst, the center frequency of each successive pulse f_m is increased by a constant frequency step Δf:

$$\{f_m = f_0 + m\Delta f | m = 0, 1, ..., M-1\}$$

where f_0 is the base frequency. The total bandwidth of the burst, i.e., M times the frequency step Δf, determines the radar range resolution. The total number of bursts, N, determines the Doppler or cross-range resolution. The returned pulse is heterodyned and quadraturely detected in the radar receiver. The received signals are measured at the pulse repetition rate at evenly spaced MN time instants

$$\{t_{mn} = (m + nM)\Delta T \,|\, m = 0, 1, ..., M-1; n = 0, 1, ...N-1\}$$

where ΔT denotes the time interval between pulses. To form a radar image, after collecting the returned signals, the M-by-N complex data are organized into a two-dimensional array, which represents the unprocessed spatial frequency signature of the object, $S(f_{mn}, t_{mn})$.

The received frequency signatures of the bursts can be treated as the time history series of the object's reflectivities at each discrete frequency. The radar processor uses the frequency signatures as the raw data to perform range and Doppler processing. Range processing functions as a matched filter for use with pulse compression, which removes frequency or phase modulation and resolves range. For stepped-frequency signals, the range processing performs M-point inverse discrete Fourier transforms (IDFT) for each of the N received frequency signatures. Therefore, N range profiles (i.e., the distribution of the object's reflectivities in range), each containing M range cells, can be obtained. For each range cell, the N range profiles constitute a new time history series, which is sampled at the baseband with N in-phase (I-channel) and N quadrature-phase (Q-channel) data. Then, the Doppler processing takes the discrete Fourier transform (DFT) for the time history series and generates an N-point Doppler spectrum, or profile. By combining the M Doppler spectra at M range cells, finally, the M-by-N image is formed. Therefore, the radar image is the object's reflectivities mapped onto the range-Doppler plane.

The range or slant-range resolution Δr_{sr} is determined by the bandwidth of the radar signal. For a stepped-frequency signal with M frequency steps of stepping frequency Δf, the range resolution is

$$\Delta r_{sr} = \frac{c}{2M\Delta f}$$

The Doppler resolution Δf_D is approximately $1/T$, where T is the imaging time duration or the observation time. The Doppler information is used to obtain the cross-range; the cross-range resolution is determined by the angle extent of the synthesized apertures during the imaging time duration, and has the form

$$\Delta r_{cr} = \frac{\lambda}{2\Omega T}$$

where λ is the wavelength of the radar signal and Ω is the rotation rate of the object. A longer observation time potentially provides higher Doppler or cross-range resolution, but causes more range and phase errors. These errors can degrade the resolution of the Doppler spectrum when the Fourier transform is used.

10.1.2   Conventional Motion-Compensation and Residual Phase Errors

Motion compensation is a very important step to achieve a clear radar image. As was mentioned earlier, because of the motion between radar and object, the process of estimating the motion and removing the extraneous phase term $\exp\{-j4\pi f_0 R(t)/c\}$ is called *focusing*. This is the first step of the motion-compensation procedure and compensates only the motion of the whole object.

In contrast with SAR, ISAR uses the object's pitch, roll, and yaw motions and measures the Doppler spectrum to generate images of the object. As is known, Doppler processing using Fourier transforms is adequate only in the case that the scatterers remain in their range cells and that their Doppler frequency shifts are constant during the entire observation time. If the scatterers drift out of their range cells or their Doppler frequency shifts are time-varying, then the constructed image becomes blurred. The complete motion-compensation procedure should be a process of establishing aligned range and stationary phase change-rate for all individual scatterers.

Conventional motion compensation is a gross compensation for the whole object. It performs mainly the range tracking and Doppler tracking. When an object is moving smoothly, conventional motion compensation is good enough to produce a clear image of the object. The conventional approach for range tracking uses a simple hot-spot tracking [17]. It is based on a single dominant scatterer and performs curve-fitting to its unwrapped phase function. Once it localizes a single dominant scatterer in the object, it relies on the phase history information of this scatterer as a reference phase to correct the whole Doppler spectrum. At each Doppler frequency, the correction is performed by multiplying the complex conjugate unwrapped phase function to the complex data at each range cell. The Doppler tracking of the object can be done by sliding-window Doppler measurement. It computes the Doppler centroid within a short time-sliding window and makes curve-fitting to the Doppler centroid curve. Then, the smoothed curve is used to generate a phase-correcting function for motion compensation.

However, when an object has a complex motion, such as pitching, yawing, rolling, or maneuvering, conventional motion compensation for the whole object is not sufficient to produce an acceptable image for viewing and analysis. There is a limit on the amount of an object's rotation that the radar imaging technique can tolerate. When an object's rotation is beyond the scope of the limit, the Doppler spectrum becomes smeared. Generally, for a large ship with slow motion, the image smearing could be minimal. However, for an aircraft with maneuvering, the image smearing could be significant.

In this case, a more sophisticated motion-compensation procedure for each individual scatterer is needed. The goal is to keep each scatterer within its range cell and to maintain constant Doppler frequency shift for each of them. Thus, the Fourier transform can be applied properly to construct a clear image of the object.

When a dominant or reference scatterer is well motion-compensated, other scatterers may still drift through their range cells. This kind of Doppler spectrum smearing may be alleviated by applying so-called polar reformatting, which attempts to eliminate other scatterers' drifts through their range cells. From (10.5) and (10.6), the spatial frequencies f_x and f_y are sinusoidal functions of the rotation angle θ. In order to take advantage of the fast Fourier transform, the frequency signature data must be sampled to produce new samples on a two-dimensional rectilinear grid, i.e., the sample points on the polar sampling grid must be conformed to the desired sample points on a rectangular sampling grid [184].

The smeared Doppler spectrum caused by large rotation angles or long observation time can also be alleviated by using the sub-aperture approach [2]. In this approach, by estimating the frequency variation with time, the size of the time - window should be chosen small enough such that there is no significant phase error during the short time interval. Thus, the Fourier transform can be applied, and sub-images can be generated without significant blurring. By calculating sub-spectra for sub-apertures, the frequency shift from one sub-aperture to the next can be found. Then, a correction function is generated from the frequency-shift function, and focusing is performed via phase correction and time-domain interpolation. This approach may combine sub-apertures coherently to achieve higher resolution.

The sub-aperture approach is basically a short-time Fourier approach and uses the time-variation property of the Doppler spectrum. Since the time window of the short-time Fourier transform (STFT) is fixed, there is a trade-off between the time window and the frequency window. High time resolution (or short time window) results in poor frequency resolution, and high frequency resolution results in poor time resolution. To enhance the resolution with the STFT, super-resolution spectrum analysis may be applied in some cases.

In the case of an object with significant maneuvering, even sophisticated motion compensation is still not sufficient and the residual of the motion is still large. With large residuals of the motion or phase errors, individual scatterers may still drift through range cells; thus, the Doppler spectrum may still be time-varying. Therefore, the resulting image becomes blurred if the Fourier transform is used.

Motion-induced time-variation in the Doppler spectrum has been used by sub-aperture algorithms, which attempt to perform time-frequency processing to resolve a blurred Fourier spectrum to a certain level of resolution. If a high-resolution time-frequency analysis can be used to replace the Fourier transform, then the restrictions of the Fourier transform can be lifted. Thus, there is no need to eliminate the drift through range cells and no need to keep Doppler frequency shift constant for each scatterer. By using high-resolution time-frequency analysis, the motion of the individual scatterer is actually examined at each time instant. Since each scatterer has its own range and its own Doppler shift at each time instant, there is no scatterer overlapping, and therefore, no image blurring occurs.

10.1.3   Motion-Induced Time-Varying Doppler Spectrum

As was mentioned earlier, the restrictions imposed by Fourier transform processing can be lifted if a processing method other than the Fourier transform is used to deal with the time-varying Doppler spectrum. From a frequency analysis point of view, the complex motion causes the Doppler frequency to be time-varying. Therefore, the time-varying Doppler frequency shift can be very well analyzed, and the image blurring caused by the time-varying Doppler spectrum can be resolved without applying sophisticated motion compensation. On the other hand, since the time-varying Doppler frequency shift can also be induced by uncompensated phase errors, the time-varying Doppler processing can also help in resolving the image blurring due to these errors.

It is necessary first to examine the relationship between motion and the time-varying spectrum before introducing joint time-frequency analysis for Doppler processing. When the object in Fig. 10–2 has complex translational and rotational motion, the phase of the returned signal from the object is

$$\Phi_r(r_t) = 2\pi f \frac{2r(t)}{c} = 2\pi f \frac{2[R(t) + x\cos\theta(t) - y\sin\theta(t)]}{c}$$

where it is assumed that $R(t) = R_0 + \upsilon_R t$ and $\theta(t) = \theta_0 + \Omega t$.

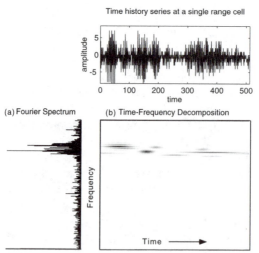

Fig. 10–4 (a) Fourier spectrum and (b) time-varying spectrum for a single range cell.

Since the time-derivative of phase is frequency, by replacing $\theta(t)$ with $\theta_0 + \Omega t$ and $R(t)$ with $R_0 + \upsilon_R t$ and taking the time-derivative of the above phase function, the Doppler frequency shift can be derived as

$$f_{D_{\text{trams}}} = \frac{2f}{c}\upsilon_R$$

and

$$f_{D_{\text{roc}}} = \frac{2f}{c}[x(-\sin\theta_0\Omega - \cos\theta_0\Omega^2 t) - y(\cos\theta_0\Omega - \sin\theta_0\Omega^2 t)]$$

where θ_0 is the initial object angle. From this equation, it is clear that when υ_R or Ω is changing with time, the Doppler frequency shift is time-varying. Even if the rotation rate Ω is constant, the rotational motion-induced Doppler frequency shift is still time-varying.

Another source of time-variation in the Doppler spectrum may result from the uncompensated phase errors due to irregularities in the object's motion, the fluctuation of the rotation rate, the fluctuation in localizing the rotation center, inaccuracy in tracking the phase history, and other variations of the system and the environment. From the relationship between the range and the phase given in (10.1), the phase is very sensitive to range variation. For example, for a wavelength of $\lambda = 3$ cm, even a 0.1 cm range error can cause 24^0 phase deviation. Since the residual phase errors may vary with time, the Doppler frequency also varies with time. As is known, the Fourier spectrum does not support the instantaneous frequency spectrum with time. By representing the time-varying Doppler spectrum with the Fourier transform, the Doppler spectrum becomes smeared, as shown in Fig. 10–4, where the Fourier transform (Fig. 10–4a) and a time-frequency transform (Fig. 10–4b) are applied to a real data time history series. Therefore, the Fourier mapping of the object's scatterers becomes blurred, while the time-frequency mapping is clear for each time instant.

Because of the time-varying property of the Doppler spectrum, an efficient way to resolve the smeared Fourier spectrum and blurred radar image is to apply a joint time-frequency transform. The joint time-frequency transform actually decomposes the phase function into instantaneous time slices. At each time slice, the Doppler frequency components are fixed, possessing the Doppler resolution provided by the joint time-frequency transform.

To achieve superior resolution and unbiased estimation of the instantaneous frequency spectrum, the time-frequency distribution series (TFDS) can be applied. The TFDS is an efficient way to achieve focusing and to resolve the image blurring problem caused by the time-varying property of Doppler spectrum. In the next sub-section, applications of the joint time-frequency transform to the range-Doppler radar imaging system to achieve superior image resolution will be discussed.

Since ISAR image construction is essentially a process of spectral analysis, super-resolution techniques of modern spectral estimation [103] can also be used to construct sharp images during a short observation time. Super-resolution does generate sharper spectral peaks than the Fourier spectrum's. However, spurious peaks can be introduced into the spectrum if the size of the autocorrelation matrix is not adequate. In noisy data situations, since the least square solution is sensitive to noise, spectrum perturbation may occur. Also, the amplitude of the peak spectrum may not be proportional to the strengths of the scatterers, and some weak scatterers may be missed. Further, super-resolution depends on the selected order and could become unstable. Compared with super-resolution, joint time-frequency analysis provides a natural way to achieve superior resolution, while the super-resolution approach simply sharpens the Fourier spectrum and generates a pseudospectrum.

10.1.4   Joint Time-Frequency Analysis for Radar Range-Doppler Imaging

Thus far, discussion has centered on the basic concept of radar range-Doppler imaging and the conventional radar imaging system based on the Fourier transform. To achieve superior resolution and unbiased estimation of the instantaneous frequency spectrum, the time-frequency properties of the joint time-frequency transform are very useful. By replacing the conventional Fourier transform with a joint time-frequency transform, a 2D range-Doppler Fourier frame becomes a 3D time-range-Doppler cube. Sampling in time yields a time sequence of 2D range-Doppler images which can be viewed [21]. Each individual time-sampled image from the cube provides not only higher resolution, but also the temporal information available within each observation time.

Next will be presented an example of the application of the TFDS to the construction of a high-resolution range-Doppler image. The objective of the simulation is to explore the potential benefit of time-frequency analysis in constructing range-Doppler images.

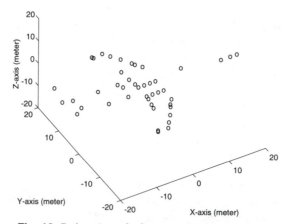

Fig. 10–5 An aircraft characterized by scatterers.

Fig. 10–5 shows a simulated aircraft in terms of its 3D reflectivity density function $\rho(x,y,z)$, which is characterized by scatterers along with their reflectivities. The aircraft is initially located at a point in three-space ($x_0 = 70$ m, $y_0 = 1600$ m, $z_0 = 100$ m). It has only translational motion with a velocity of 120m/s along a track at $130°$ from the u-axis in the global coordinate (u,v) in Fig. 10–2.

The translational motion can induce an equivalent angular rotation of the aircraft with an angular velocity of 4.3°/s.

In this simulation, the radar is assumed to be operating at 9000 MHz. A total of 64 stepped frequencies are used in each burst. To achieve 1.0 m range resolution, a total of 150 MHz bandwidth is required, and, therefore, the stepping frequency is 2.34 MHz. The pulse repetition frequency used in the simulation is 20,000 pulses/s, which is high enough to cover the entire aircraft. The observation time should be long enough to achieve the desired resolution. Here, 0.82 s observation time, or 256 samples of the time history series, is assumed. Therefore, the radar image consists of 64 range-cells and 256 Doppler frequencies. For the above initial kinematic parameters of the simulated aircraft, the cross-range resolution is 0.44 m.

A fluctuation in the velocity

$$v(t) = v(0) + \Delta v(t)$$

can induce an equivalent radial velocity fluctuation as well as an equivalent angular velocity fluctuation. It is interesting to introduce a velocity fluctuation by assuming that $\Delta v(t)$ is a sinusoidal function of time. Even if the fluctuation causes a maximum velocity variation of only 0.5m/s with the velocity specified above (which results in a maximum range fluctuation of 0.1 m), the uncompensated phase error due to this fluctuation can cause the constructed image to be blurred, as shown in Fig. 10–6a, where range and Doppler tracking have been applied.

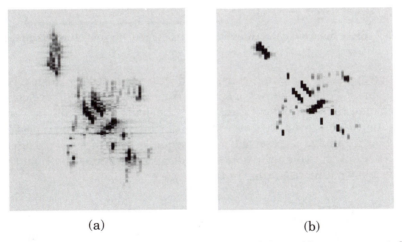

(a)   (b)

Fig. 10–6 Radar image of the aircraft from simulated data with uncompensated phase errors by using (a) Fourier transform and (b) joint time-frequency transform.

Clearly, the uncompensated phase error causes the Doppler spectrum to vary with time. Dealing with the time-varying Doppler spectrum using conventional Fourier transform processing blurs the image. If the Fourier transform is replaced with a joint time-frequency transform implemented by the TFDS, as illustrated in Fig. 10–7, a single image frame becomes a sequence of temporal image frames. In other words, the joint time-frequency transform resolves the single image frame into a stack of its temporal frame elements. Each temporal frame element's range-Doppler resolution is higher than that of the single Fourier frame.

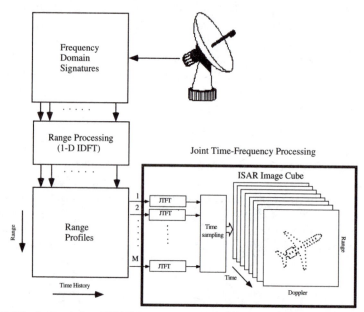

Fig. 10–7 Block diagram of ISAR image construction using joint time-frequency transform.

For comparison, the image constructed with the Fourier transform is shown in Fig. 10–6a. Fig. 10–6b shows one frame from the image sequence constructed with the joint time-frequency transform. The joint time-frequency transform yields a much improved representation of the time-varying spectrum: the blurred Fourier image is resolved into a sequence of time-varying images; the sequence not only has higher resolution, but also illustrates the Doppler changing and range walking from time to time.

10.1.5   Summary

Conventional radar imaging uses the Fourier transform, which assumes that the contents of Doppler frequencies are not changing during imaging time. However, in most cases, this assumption is not true. Due to an object's motion and rotation, the motion-induced Doppler frequency shift is actually time-varying. The time-varying Doppler spectrum can also result from uncompensated phase errors due to an object's irregular motion and inaccuracy in tracking the phase history. Therefore, by representing the time-varying Doppler spectrum with the Fourier spectrum, the Doppler spectrum becomes smeared.

To represent time-varying spectra, a joint time-frequency transform is desirable. The joint time-frequency transform can very well represent the local time-frequency structures of the radar data. Replacing the conventional Fourier transform with a joint time-frequency transform converts a 2D range-Doppler Fourier frame into a 3D time-range-Doppler cube. By sampling in time, a time sequence of 2D range-Doppler images can be viewed. Each individual time-sampled image from the cube provides not only higher resolution, but also the temporal information associated with each imaging time.

10.2 Backscattering Feature Extraction[*]

ISAR imaging has long been used by the microwave radar community for signature diagnostic and target identification applications. ISAR is a very robust process for mapping the point scatterers on the target. For instance, in a typical signature diagnostic application, a target is mounted on a pedestal in a radar anechoic chamber and irradiated with an electromagnetic plane wave. The backscattered wave is then sampled over multiple frequencies and viewing angles (by rotating the target on the pedestal). By processing these samples via the ISAR imaging algorithm, the ISAR image of the target is formed that usually consists of a collection of discrete point scattering features, or scattering centers. These scattering centers on the target can be identified in the image plane as a function of down-range and cross-range and are useful for pinpointing those elements on the target which are the dominant contributors to the radar cross section (RCS) of a target in signature design applications. Similarly, in target identification scenarios, the ISAR image is acquired by tracking a maneuvering target over time using a stationary radar, in essence gathering multi-look data on the target. The formed image is then compared against existing templates in order to identify the unknown target.

While simple targets can be modeled as a collection of point scatterers, real-world targets contain many non-point scattering mechanisms. For instance, aircraft signatures consist of not only scattering centers, but also resonances from sub-skin line features such as inlets, cockpits, and antenna windows. These scattering mechanisms appear in an ISAR image as blurred clouds which extend down-range and often do not correspond to the spatial features on the target. On the other hand, the same mechanisms do provide important features which, if properly interpreted, can become important factors in the target classification process. In this second part of the chapter, we shall discuss the artifacts in the ISAR image associated with the non-point scattering physics of realistic targets and ways to extract them using joint time-frequency techniques. This section is

[*] Contributed by Luiz C. Trintinalia and Hao Ling, Department of Electrical and Computer Engineering, University of Texas, Austin, Texas 78712-1084. This work was supported by the Joint Services Electronics Program under Contract No. AFOSR F49620-95-C-0045 and in part by the Air Force Wright Laboratory through DEMACO subcontract DEM-95-UTA-55. "The United States Government is authorized to reproduce and distribute reprints for governmental purposes notwithstanding any copyright notation hereon."

organized as follows: We first give a brief overview of the ISAR imaging principle and show how non-point scattering mechanisms can create artifacts in the image. We then describe how time-frequency processing, in particular the adaptive spectrogram, can be applied to remove these non-point scattering features from the ISAR image. Finally, we present several processing examples, using data from numerical simulation and actual chamber measurement, on how this processing can lead to a cleaned image and how the non-point scattering features can be displayed in an alternative feature space to provide additional information on the target feature.

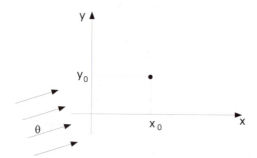

Fig. 10–8 Ideal scattering center excited by plane wave.

10.2.1   ISAR Algorithm and Effects of Non-Point Scattering Mechanisms on ISAR Imagery

Inverse synthetic aperture radar imaging provides a way to map data gathered at multiple frequencies and aspect angles into a two-dimensional image (the ISAR image) whose points correspond to x-y coordinates in space (called *down-range* and *cross-range*, respectively). For a target containing an ideal scattering center with strength A at the position (x_0, y_0), when excited by an electromagnetic wave incident at an angle θ and frequency f (see Fig. 10–8), the resulting scattered field can be written as

$$S(f, \theta) = A\exp(-j2k_x x_0 - j2k_y y_0) \tag{10.7}$$

where

$$k_x = \frac{2\pi f}{c}\cos\theta, \quad k_y = \frac{2\pi f}{c}\sin\theta$$

If we take the inverse Fourier transform of (10.7) with respect to the variables $2k_x$ and $2k_y$, we obtain

$$s(x, y) = A\delta(x - x_0)\delta(y - y_0) \tag{10.8}$$

This is the essential idea behind the ISAR processing: ideal scattering centers correspond to ideal points in the ISAR image. Of course, due to limited bandwidth and aspect angle, we will see a spot of finite size, rather than a point, in the image.

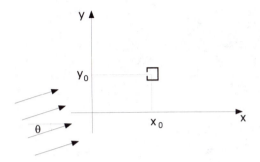

Fig. 10–9 Small cavity excited by plane wave.

Real targets, however, cannot always be suitably represented by a collection of scattering centers. Cavities, ducts, and other sub-skinline structures present in real targets produce scattered signals that have very strong frequency dependence and cannot be well focused in the ISAR plane. Consider, for example, the small cavity depicted in Fig. 10–9. Disregarding the scattering contribution from its exterior and assuming, for simplicity, that the energy coupled through its small aperture will be isotropically radiated at a single frequency f_0 (one of the resonance frequencies of the cavity), we can write the expression for this scattered signal as

$$S(f,\theta) = A \frac{\exp\left(-j2k_x x_0 - j2k_y y_0\right)}{\alpha + j\,2\pi\left(f - f_0\right)} \tag{10.9}$$

In deriving this expression, we assumed that a damping resonance frequency (with damping factor α) starts to radiate from the aperture (x_0, y_0) when the incident wave reaches the cavity.

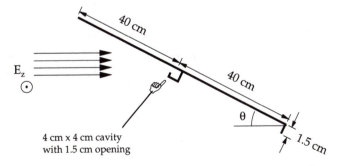

Fig. 10–10 Conducting strip with small cavity and fin.

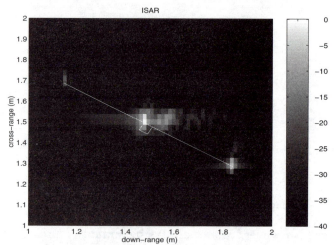

Fig. 10–11 ISAR image obtained for the geometry shown in Fig. 10–10. Simulation was carried out for f = 5-15 GHz and θ = 0-60°.

If we further assume that data are gathered over a rectangular window in the k-plane along the k_x axis, having a small angular aperture

$$k_x \cong k \qquad k_y \cong k\theta$$

but large bandwidth ($\Delta k_x >> \Delta k_y$), the corresponding expression in the ISAR plane can be shown to be

$$s(x, y) \cong B\mathrm{sinc}[\Delta k_y (y - y_0)]\exp(j2k_0 x)\exp(-2\alpha x/c)u(x - x_0) \qquad (10.10)$$

We notice that the ISAR feature of this signal, while still localized around the cross-range of the cavity aperture, spreads out in the down-range direction. As an example, consider the simple structure of a strip containing an open cavity, shown in Fig. 10–10. We generated the scattered data using a standard two-dimensional moment method simulation. The simulation was carried out for frequencies from 5 to 15 GHz, with an angular aperture of 60°. Fig. 10–11 shows the obtained ISAR image for this target, with the outline of the target superimposed on it to facilitate the identification of the scattering mechanisms. As expected, three scattering centers, corresponding to the left and right edge of the strip and the cavity exterior, can be identified in the image. Besides these scattering centers, we also see a large cloud, spreading through the down-range, which corresponds to the energy coupled inside the cavity, and re-radiated at its corresponding resonance. This cloud not only makes the image more crowded, but may in some cases obscure some other important scattering centers. This portion of the scattered signal contains information on the resonance frequency of the cavity.

10.2.2 Time-Frequency Processing of ISAR Images

For signals that contain both scattering-center mechanisms and resonances, neither the time domain nor the frequency domain analysis will be able to reveal all the features present in it. In such cases, the time-frequency plane is the only display that will allow us to discriminate these two phenomena (as well as other more complex dispersive mechanisms). The short-time Fourier transform (STFT) is the most standard procedure and a simple to obtain a time-frequency display. This processing technique involves the use of a sliding window in the time domain. For each position of this window, a Fourier transform of the windowed data is generated, and therefore we can obtain an image (called the *spectrogram*) showing the spectrum of the analyzed signal for each position of the window. By using this technique, we obtained a time-frequency display of the backscattered data for the geometry depicted in Fig. 10–10, at a single angle of $\theta = 25°$. In this display, shown in Fig. 10–12, we can identify the scattering centers as vertical lines and the resonances as horizontal lines.

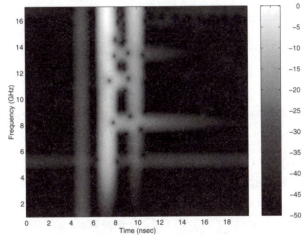

Fig. 10–12 STFT of backscattered signal from the geometry shown in Fig. 10–10 for θ = 25° and f = 50 MHz το 18 GHz.

It would be interesting then to combine this time-frequency processing with the ISAR imaging technique so that we can still pinpoint the positions of scattering centers in space and the positions of the resonances in the frequency domain. To accomplish that, we process each line (cross-range) of the obtained ISAR image using the STFT engine, thus generating a third axis for this display. Since we will be sliding our window along the down-range, this new axis will be proportional to k_x, which for a small angular aperture is proportional to the frequency. This new display has three dimensions: down-range, cross-range, and frequency. Each frequency slice of this "cube" will be a narrow-bandwidth ISAR image, and so it will allow us to identify scattering centers, which will be present in all slices, and resonances, which will appear only in the slices close to the resonance frequencies. Fig. 10–13 shows such a display, obtained for the same example referred to earlier. It is possible now to identify the three scattering centers, and also to estimate the resonance frequencies present in the signal (which show up as large clouds in some slices). However, even though we gained some insights with this new display, the frequency resolution came at the cost of losing down-range resolution.

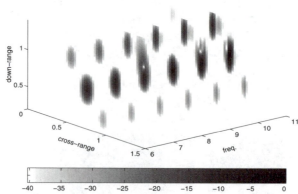

Fig. 10–13 Joint time-frequency ISAR display obtained by applying the STFT to the ISAR image in Fig. 10–11.

Clearly, a different processing engine is needed other than the STFT to gain frequency resolution, without sacrificing the down-range resolution. This engine is the adaptive Gabor representation (AGR), to be described next.

10.2.3   Adaptive Joint Time-Frequency ISAR

The adaptive Gabor representation was introduced in the signal processing community by Qian et al (see [146] and [151]). Since the rigorous treatment of this algorithm can already be found in Chapter 8, only those aspects that are closely related to our application are described here. The basic idea of this procedure is to expand a signal $s(t)$ in terms of a set of localized Gaussian functions $h_p(t)$ with an adjustable standard deviation σ_p and a time-frequency center (t_p, f_p):

$$s(t) = \sum_{p=1}^{\infty} B_p h_p(t) \tag{10.11}$$

where

$$h_p(t) = (\pi\sigma_p^2)^{-0.25} \exp\left\{-\frac{(t-t_p)^2}{2\sigma_p^2}\right\} \exp\{j2\pi f_p t\} \tag{10.12}$$

The coefficients B_p are found one at a time by an iterative search procedure. One begins at the stage $p = 1$ by choosing the parameters σ_p, t_p, and f_p such that $h_p(t)$

is most "similar" to $s(t)$, i.e., for which the inner product between $s(t)$ and $h_p(t)$ is the largest. The next B_p is found using the same procedure after the orthogonal projection of $s(t)$ onto $h_p(t)$ has been removed from the signal. This procedure is iterated to generate as many coefficients as needed to accurately represent the original signal.

Several comments are in order. First, it can be shown that the norm of the residue monotonically decreases and converges to zero. Therefore, adding a new term in the series does not affect the previously selected parameters. Second, because this representation is adaptive, it will be generally concentrated in a very small sub-space. As a result, we can use a finite set of elementary function $\{h_p(t)\}$ to approximate the signal, with a residual error as small as one wishes. Also, since random noise in general is distributed uniformly in the entire space, this sub-space representation actually increases the signal-to-noise ratio. Finally, the major difficulty in implementing this algorithm is the determination of the optimal elementary function at each stage. In our implementation, we used the same guidelines as those given in [151].

The adaptive spectrogram has two distinct advantages over conventional time-frequency techniques such as the short-time Fourier transform. First, it is a parameterization procedure that results in very high time-frequency resolution. More importantly for our application, this representation allows us to automatically distinguish the frequency-dependent events from the frequency-independent ones through the extent of the basis functions. From expression (10.12), it can be seen that scattering centers, i.e., signals with very narrow length in time, will be well represented by basis functions with very small σ_p. Frequency resonances, on the other hand, will be better depicted by large σ_p.

We can now replace the STFT engine by the adaptive Gabor representation in the procedure described in 10.2.2. In our implementation, we run the adaptive Gaussian algorithm for each line (cross-range) of the ISAR image, thus extracting a set of Gaussian basis functions that adequately represent that image. Since the data are in effect parameterized using the adaptive Gabor representation, we no longer have to create the three-dimensional display, like that shown in Fig. 10–13, using the STFT. This saves us from having to store a very large three-dimensional image.

Once we obtain the parameterized representation of the data, we can distinguish the scattering-center mechanisms from the resonance information based on the variance of the time-frequency basis. Those with small variances correspond to the scattering-center mechanisms. Using only the Gaussians in this set, we can reconstruct our ISAR image (by simply adding up the selected basis functions), and thus generate a clean, enhanced image that contains only scattering-center information. We can also reconstruct the ISAR image using only the large-variance Gaussians, those related to the resonances and other discrete frequency events. It is, however, more meaningful to view these frequency-dependent mechanisms in the Fourier transform domain of the ISAR image. By doing so, we will obtain a dual frequency-aspect display where resonances and other frequency-dependent mechanisms can be better identified. The procedure is illustrated graphically in Fig. 10–14.

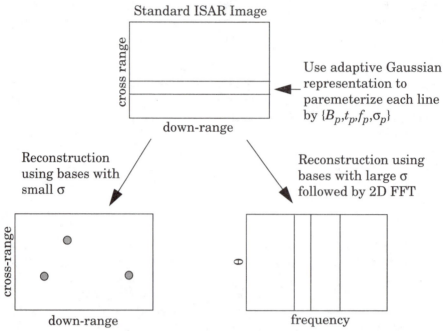

Fig. 10–14 By applying the adaptive joint time-frequency procedure to each cross-range line of the standard ISAR image, we can separate the original image into two new images based on the width of the Gaussian basis function: an enhanced ISAR image containing only scattering centers and a frequency-aspect image containing only resonance information.

10.2.4   Examples

We shall now present two examples of the adaptive joint time-frequency ISAR processing. The first example is based on numerically simulated data for a perfectly conducting plate containing a long duct open at one end, shown in Fig. 10–15. Fig. 10–16a shows its ISAR image at 40° from edge on. The scattering data were simulated numerically using a method of moment electromagnetics code. We notice in the image that, in addition to the three scattering centers corresponding to the two edges of the plate and the mouth of the duct, there is also a very strong return outside of the target. This return corresponds to the energy that gets coupled into the duct, travels down the length of the duct, hits the end and reflects back to the duct mouth and is finally re-radiated towards the radar. Fig. 10–16b shows the enhanced ISAR image of Fig. 10–16a, obtained by applying the above algorithm and keeping only the small-variance Gaussians. We see that only the scattering-center part of the original signal remains in the image, as expected. Fig. 10–16c shows the frequency-aspect display of the high-variance Gaussians. We see three equispaced vertical lines. They correspond to the cut-off frequencies of the waveguide modes excited in the duct, which should occur at 3.75, 7.5, and 11.25 GHz based on the transverse dimension of the duct. Indeed, we see that they occur close to these frequencies and are almost aspect independent.

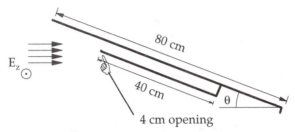

Fig. 10–15 Conducting strip with a long cavity.

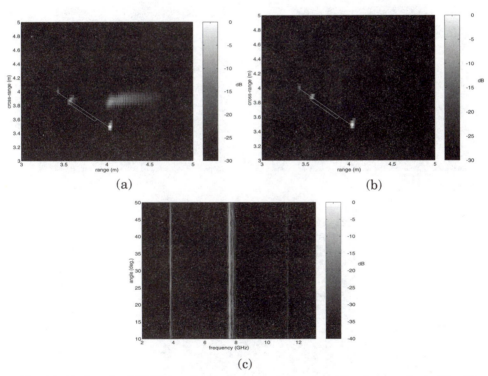

Fig. 10–16 Standard ISAR image (a) of conducting strip with a long cavity in Fig. 10–15, obtained for f = 2-13 GHz, θ = 25°-55°. Enhanced ISAR image (b) and frequency aspect display (c) obtained by applying AGR to the ISAR image in (a).

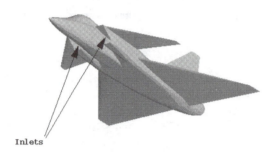

Inlets

Fig. 10–17 VFY-218 model

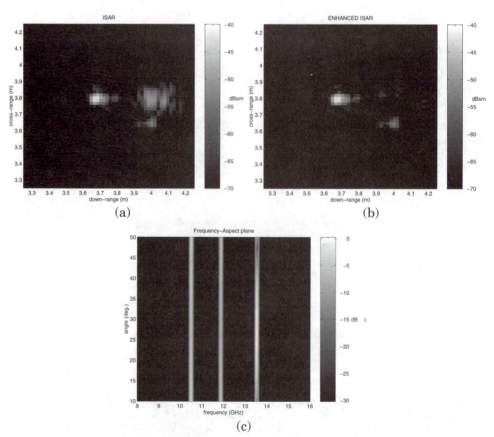

(a)   (b)

(c)

Fig. 10–18 Standard ISAR image for VFY-218 model in Fig. 10–17a obtained for $f = 8$ - 16 GHz, $\theta = 0° - 40°$. Enhanced ISAR image (b) and frequency aspect display (c) obtained by applying AGR to the ISAR image in (a).

The second example uses the chamber measurement data of a 1:30 scale model Lockheed VFY-218 airplane provided by the EMCC (Electromagnetic Code Consortium) [182]. The airplane has two long engine inlet ducts, as shown in Fig. 10–17, which are rectangular at the open ends, but merge together into one circular section before reaching a single-compressor face. As we can clearly see in the conventional ISAR image of Fig. 10–18a for the vertical polarization at 20° near nose-on, the large cloud outside of the airframe structure is the inlet return. Fig. 10–18b shows the enhanced ISAR image of Fig. 10–18a, obtained by applying the above joint time-frequency ISAR algorithm and keeping only the small-variance Gaussians. We see that only the scattering-center part of the original signal remains in the image, as expected. Notice that the strong return due to engine inlet has been removed, but the scattering from the tail fin remains. Fig. 10–18c shows the frequency-aspect display of the high-variance Gaussians. A number of equispaced vertical lines can be seen between 10.5 and 13.5 GHz. Given the dimension of the rectangular inlet opening, we estimate that these frequencies correspond approximately to the cutoffs of the second and higher modes in the waveguide-like inlet.

10.2.5 Summary

In this section, we presented an example of how the adaptive spectrogram can be applied to the range axis of the ISAR image to process data from complex targets containing not only scattering centers but also other frequency-dependent scattering mechanisms. This application is the exact dual of what was described in the first part of this chapter, where the joint time-frequency technique is applied to the cross-range (or the Doppler frequency) axis of the ISAR image for motion compensation. The adaptive spectrogram is well suited for the present application not only because of its robustness and high resolution, but because the frequency extent of the resulting Gaussian basis functions allows the natural separation of the scattering mechanisms into target scattering centers and resonances. The computational cost of the adaptive search algorithm is high. However, once obtained, the parameterized feature set can be used to very quickly reconstruct the ISAR image. The adaptive joint time-frequency ISAR algorithm allows the enhancement of the ISAR image by eliminating non-point scatterer signals, thus leading to a much cleaner ISAR image. It also provides information on the extracted frequency-dependent mechanisms such as resonances and frequency dispersions. This is accomplished without any loss of resolution.

CHAPTER 11

Localization of Brain Functions by Joint Time-Frequency Representations[*]

The human brain is often considered to be the most complex biological structure to exist. It consists of large numbers of neurons and interconnections which process and transmit information. Understanding the working brain and nervous system has been an active subject of research across a wide variety of disciplines. Although investigations by life scientists on biological and biochemical issues concerning brain function are important, mathematical models and engineering approaches also play useful roles in the quest for understanding the brain. In recent years, technological advances have allowed functional neuroimaging to be performed, through functional MRI and positron emission tomography (PET), which measure changes in blood flow, and through magnetoencephalography (MEG) and electroencephalography (EEG), which measure the magnetic and electrical fields. Both MEG and EEG provide excellent temporal resolution, with the EEG being far less costly. Currently, the EEG and its variant, the event-related potentials (ERPs), are the most widely used methods of assessing brain function.

Our investigation is related to a problem of major significance, in both basic and clinical neuroscience: the determination of the location of functional areas related to the generation of electrical activity in the human brain. The solution of

[*] Contributed by Mingui Sun and Robert J. Sclabassi, Department of Neurosurgery, University of Pittsburgh, Pittsburgh, PA 15213. This work was supported by the National Institute of Health Grant No. MH41712-06, and the Biomedical Engineering Grant Program of the Whitaker Foundation.

this problem will provide the capability to answer important questions, in a non-invasive means, concerning activated areas of the brain where information is processed. In addition, the identification of active neural substrates will aid in making critical neurosurgical decisions.

In this chapter, we first provide a biological background about the nervous system, its function, and the contribution of the neurons to the observed EEG. Then, signal processing methods applied to the EEG are briefly discussed, with particular emphasis on time-frequency analysis. We then briefly introduce techniques for source localization based on mathematical models of a volume conductor and a dipole current source, formulated as the inverse problem of the EEG. Next, our approach to the problem of noise-suppression in the multichannel EEG is described using time-frequency analysis and synthesis techniques. In this approach, we compute the time-frequency distribution series (TFDS) for each channel of the EEG and the result is weighted and averaged. This process suppresses cross-terms in the individual TFDS and produces a single time-frequency distribution. Regions of interest which contain the desired signals are specified on this distribution to exclude the undesired components of the EEG from the localization process. The signal of interest is reconstructed using time-frequency synthesis and the source of the functional activity in the brain is estimated and mapped numerically by solving a least-squares problem. Finally, we apply this method to localize sleep spindle data, an EEG pattern observed during sleep.

11.1 Biological Background of the Brain

The individual nerve cell, or neuron, is the basic functional unit of the nervous system involved in the transmission of information [100]. The neuron consists of three parts: 1) a dendritic branching through which input information is transferred to the cell, 2) a body (or soma) which serves to integrate this information, and 3) an axon, which is a segment transferring information to other neurons. Each neuron is in contact through its axon and dendrites with other neurons, so that each neuron is an interconnecting segment in the network of the nervous system.

A synapse is a specialized site of contact between neurons. The axon of a neuron may terminate in only a few or in many thousands of synapses. A thousand synaptic terminals per axon has been estimated as a rough average. The dendrites and soma of a single neuron may receive synaptic contact from several hundred to over 15,000 axons. Within the central nervous system, nerve cells are highly interconnected. It has been estimated that there are 10^{10} neurons within the brain of man, and that the number of synaptic junctions approaches 10^{14}. The number of possible permutations of neuronal interconnections in even a small region of the central nervous system is very large indeed.

The brain, in which the longest distance between any pair of points is approximately 17 cm from the front to back, integrates neuronal functional units forming topological regions. Within these regions exist compact groups of neurons, called *nuclei*, which are often anatomically identifiable. Tracts of axons connecting these nuclei can be traced from region to region, and it is to such relatively complex nuclear regions that the various functions of the nervous system are related, and which are the putative sites of the generators for electrical activity observed on the scalp. When the brain is involved in a certain task, corresponding nuclei are activated and information is processed and transmitted. The global effects of these activities may be observed as a non-stationary signal, which has been termed the electroencephalogram (EEG), and is thought to be the synchronized sub-threshold dendritic potentials of many neurons summed. The EEG is normally recorded from the human scalp, and typically has amplitudes from 10 to 100 μV and a frequency content from 0.5 to 40 Hz. Signals of 10 to 30 μV are considered low amplitude and potentials of 80 to 100 μV are considered high amplitude. The spectrum of the EEG is divided into four dominant frequency bands: δ-band (0-4 Hz), θ-band (4-8 Hz), α-band (8-13 Hz), and β-band (13-30 Hz).

11.2 The Need for Signal Processing

The EEG was first discovered by Berger in 1924, who was thus the first to open a window on the functional world of the brain. However, what he observed was "noisy" and "chaotic" waves. Since then, considerable research has been conducted to explain these mysterious waves. Fig. 11–1 shows a segment of the EEG recorded from a male adult during sleep, where it can be observed that the waveforms are quite complex in structure and their morphological shapes vary sharply with a short period of time. Traditionally, the EEG is read by clinicians whose interpretations are based on their experience. As a result, uncertainty is involved due to the level of training received and the form of graphics presented. In recent years, computer signal processing has gained popularity in the EEG community (see [66] and [162]). The spectral characteristics of the EEG can now be evaluated efficiently using the discrete Fourier transform. This method assumes that the observed data are stationary over short periods of time (on the order of 2 to 4 s). Under this assumption, the frequency band classification referenced above, which summarizes the spectral content of the EEG, provides a convenient basis for data analysis and comparison; however, this classification is not effective when the EEG exhibits significant non-stationarities. Recently, time-frequency analysis techniques (see [32] and [73]) have been applied to the EEG and other related measures such as event-related potentials (ERPs) (see [56], [162], and [163]). These techniques provide a new perspective in which the time-dependent spectrum of the EEG can be observed. Earlier applications were dominated by the spectrogram, which was based on the windowed short-time Fourier transform [104]; later the Wigner-Ville, Page, and Rihazcek distributions were applied to the EEG and ERPs (see [56], [102], [162], [163], and [183]). After extensive studies on the general class of distributions initialized by Cohen [26], the Choi-Williams distribution was developed and applied to the EEG and electrocorticogram [189]. In addition, the cross time-frequency distributions which measure the time-frequency coherence of a pair of signals were used for analyzing multichannel EEG (see [162] and [163]) and for identifying epileptic seizures [189].

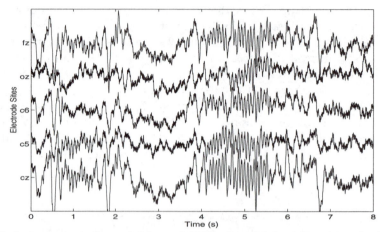

Fig. 11–1 A segment of a multichannel EEG of an adult male subject during sleep. Only five channels are shown out of a total of 64 channels recorded. This EEG segment belongs to Stage 2 Quiet Sleep, where a sleep spindle is present between the fourth and seventh seconds.

Most previous applications of signal processing to the EEG have focused on finding temporal and spectral properties of the signal such that hypotheses relating to changes of these measures with respect to different physiological and pathological conditions can be tested. In this and other investigations (see [44], [62], [101], and [129]), in contrast, the geometric space of the head is the domain of interest. We are focused on the non-hypothesis-driven issue of localizing the neuronal substrates in the brain responsible for the generation of the observed patterns in the multichannel EEG (see [171] and [172]). This approach requires not only the development of mathematical models for the volume conductor of the head and optimization algorithm to solve the source parameters (to be discussed), but also sophisticated signal processing techniques to extract the desired signal from various noise interferences. As indicated previously, the EEG is a complex signal reflecting an extremely large number of microscopic neuronal activities. Some of these activities are related to the event of interest, while the rest may be considered to be noise interference. Besides this undesired neurological interference, the EEG is also contaminated by other biological and environmental noise as well, such as muscle activity, the electrocardiogram, 60 Hz components from the power line, variation in electrode contact impedance, and electromagnetic emissions. The solutions to the source localization problem could be meaningless without filtering the EEG. Traditional bandpass filtering may be applied when the signal and noise do not share the same frequency bands; however, this is seldom the situation. Time-frequency analysis techniques provide powerful alternatives for isolating the signal of interest from the contaminating noise, since noise can be identified much more easily in the joint time-frequency domain than in either the time or frequency domain alone. While noise tends to spread widely in the time-frequency plane, the signal is often concentrated, though its time-frequency profile may be complex in shape. This suggests a filtering strategy to transform the EEG into the time-frequency domain, isolate the patterns that reflect the event of interest, and then reconstruct the signal using these patterns by time-frequency synthesis.

11.3   Joint Time-Frequency Analysis and Synthesis

In order to localize the event of interest within the brain, an array of electrodes is affixed to the scalp to record the signals emitted from the source. A larger array utilized in this process usually results in more precise and stable localization. In our experiments, 64 channels of the EEG are recorded, which are processed to isolate the signal from noise. Due to the limitation of visualizing this large data set, we compute the time-frequency distribution for each channel and then average the results by

$$P(t, \omega) = \sum_{c=1}^{N_c} \omega_c P_c(t, \omega) \tag{11.1}$$

where N_c is the number of channels, and ω_c is the weight, $\omega_c = \hat{S}_C / S_c$, with $\hat{S}_C$ and S_c being, the energy within the frequency band of interest and the total energy for the signal in channel c, respectively. This averaging produces a single distribution allowing a pre-discrimination of the noise by emphasizing the channels where the signal is strong.   In addition, the cross-terms present in each individual time-frequency distribution are reduced because the noise components are often less correlated among the EEG channels than the signal components. The phase differences in the noise may also help to cancel the cross-terms when they are highly oscillatory in the time-frequency plane. In order to estimate the energy ratio ω_c, we apply a bandpass filter to all channels of the EEG with a pair of fixed cut-offs which cover all possible frequencies of the signal. The values of these cut-offs may be obtained from studies in neurophysiology.

   The selection of the time-frequency distribution $P_c(t,\omega)$ in (11.1) is rather crucial. To identify the event of interest, $P_c(t,\omega)$ should have a high resolution in the joint time and frequency domain. Therefore, we chose the time-frequency distribution series (TFDS) (also known as the Gabor spectrogram), described in Chapter 7, to compute $P_c(t,\omega)$. By adjusting the order of the TFDS, the trade-off between the resolution and the cross-term interference can be compromised effectively. In our case, this order was chosen between two and four. Fig. 11–2 shows the average TFDS for the signal in Fig. 11–1.

   Once the average TFDS is obtained, it is represented in an image form in which the regions of interest (ROI) are specified. The ROI identifies the boundaries of the time-frequency components of interest by incorporating a priori knowledge, obtained from neurophysiological studies, into the subsequent signal processing. The coordinates of the boundary are converted into a binary mask (Fig. 11–2b). This mask is then used to select the ROI in each channel of the time-frequency distribution for the reconstruction of the noise-free signal.

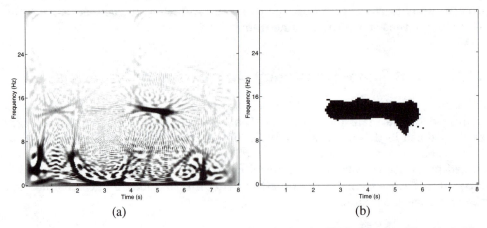

(a)   (b)

Fig. 11–2   (a) The weighted average of 64 TFDSs for the EEG shown in Fig. 11–1. The sleep spindle signal can be clearly observed in the high-frequency plane. (b) The region of interest selecting the sleep spindle signal in (a).

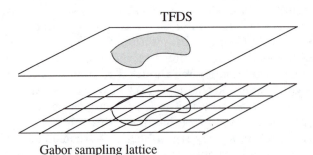

Fig. 11–3   The ROI is first identified in the highly redundant TFDS image and then mapped into the corresponding Gabor sampling lattice.

The process of constructing a time-domain signal from a modified time-frequency distribution is known as *time-frequency* synthesis. Methods of time-frequency synthesis based on quadratic distributions have been extensively studied (see [15], [42], [108], and [144]). Generally speaking, features of the signal can be visualized more easily from a quadratic time-frequency distribution, such as the STFT spectrogram, Wigner-Ville distribution, or TFDS, than from a linear representation, such as the Gabor or wavelet transform; however, quadratic distributions are highly redundant in terms of the large data size compared to the original signal. As a result, signal synthesis based on quadratic functions requires a high computational cost, which represents a significant problem in our case since, for each data segment, 64 channels of the EEG must be processed.

In order to solve this problem, we take advantage of the close relationship between the TFDS and the Gabor transform (discussed in Chapter 3). We use the TFDS to specify the ROI in the time-frequency plane where the features of the signal can be observed in high resolution. We then project the resulting ROI onto the Gabor sampling lattice to mask the Gabor coefficients, as shown in Fig. 11–3. This process eventually forms a linear representation of the time-frequency filtered signal after a few interactions (see Chapter 9 for detail). Since the Gabor coefficients are used twice in the computation of the TFDS and signal reconstruction, and the Gabor transform is a linear representation, this analysis-synthesis method provides an ideal vehicle to carry the signal between the time and time-frequency domains in this EEG application.

11.4 Model-Based Source Localization

In the source localization problem, the geometric space of the head is the domain of interest. The values of the filtered EEG signals

$$\{ \upsilon_c(t_0) \}$$

for $c = 1, 2,...,N_c$, at a selected time t_0, provide the information about the dipole location $r = (x, y, z)$ and dipole moment $m = (m_x, m_y, m_z)$ at t_0. The dipole model is valid when the actual source of functional activity is a set of densely packed, activated neurons.

Solving for the source parameters from the observed EEG has been formulated as *the inverse problem* [101]. Theoretically, a system has an inverse if and only if its input/output relationship is unique; however, this is not the case here since two different sources may produce identical waveforms at all recording channels. Therefore, this inverse problem is generally ill posed. In addition, a model of the head, including its geometry and conductivity, must be utilized to solve the source parameters r and m. The simplest model of the head includes a spherical volume conductor with shells of isotropic conductance [11]. Other complex models employ more than one dipole, more realistic head shapes, and non-homogeneous conductivity values (see [33] and [63]) which are more precise, yet more computationally expensive.

The relationship between the current source and the potential value can be formulated as the following general form:

$$\nabla[\sigma(\gamma)\nabla V(\gamma)] = s(\gamma) \tag{11.2}$$

for $r \in \Omega$ and

$$\left.\frac{\partial V}{\partial n}\right|_B = 0$$

where Ω denotes the geometric space of the head, B the boundary of Ω, $V(\gamma)$ the potential in response to the source at location γ, $\sigma(\gamma)$ the conductivity tensor (generally anisotropic), $s(\gamma)$ the current source function, and ∇ the gradient operator.

$$\nabla = \left(\frac{\partial}{\partial x}, \frac{\partial}{\partial y}, \frac{\partial}{\partial z}\right)$$

When σ and s are known and the boundary condition on B is imposed, V may be solved analytically for some simple geometric shapes, such as a layered sphere-like volume conductor excited by a dipolar current source [130]. In more realistic models, an analytic solution does not exist and the finite boundary or finite element method must be used.

Once the relationship between the source and the scalp potential is established, a numerical optimization is applied with respect to the six parameters in r and m and the measured EEG. Assuming the dipole can be translated and rotated within the head volume conductor without constraints, the optimization process minimizes the following sum of the squared errors (usually called the *residual variance*) over all channels of the EEG:

$$R = \frac{\sum_{c=1}^{N_c}\left[V_c - \upsilon_c\right]^2}{\sum_{c=1}^{N_c}\upsilon_c^2}$$

where V_c and υ_c are, respectively, the theoretical and actually measured potentials for channel c, and N_c is the number of channels. This optimization process may be carried out by a variety of algorithms, such as the simplex, steepest descent, and Newton-Raphson algorithms.

11.5 Experimental Results

In this experiment, the time-frequency filtering method is used to extract patterns of the sleep spindle signal from the multichannel EEG and localize these patterns. Previous studies have shown that the EEG waveforms are remarkably different when a person is alert and asleep. An alert person displays a low amplitude EEG of mixed frequencies in the 8 to 13 Hz range, while a relaxed person produces large amounts of sinusoidal waves (with a narrow bandwidth between 8 to 13 Hz), which are particularly prominent at the back of the head. Three alert-sleep states have been distinguished: waking (W), quiet sleep (QS), and rapid eye movement (REM). In the adult human, QS is further differentiated into four stages on the basis of brain, muscle, and eye activity. QS, REM, and occasional momentary wakings occur in a periodic sequence throughout the night, taking approximately 90 min in the adult human.

As an individual goes to sleep, α-activity is replaced by a lower amplitude, mixed-frequency voltage (Stage 1 QS), which within minutes has superimposed 1- to 2-s bursts of 12- to 14-Hz activity called *sleep spindles* (Stage 2 QS) (middle portion in Fig. 11–1), the activity investigated in this experiment. Several minutes later, high-amplitude slow waves (0.5 to 3 Hz) appear and mark the onset of Stage 3 QS. After about 10 min, these slow waves dominate the EEG and the deepest stage of sleep, Stage 4, is reached. After a return through these stages, REM sleep occurs, approximately 90 min after sleep onset.

Previous studies [67] have shown that sleep Stage 2 in which sleep spindles appear occupies more than 50% of the total overnight sleep time in a typical adult. Animal studies (see [166] and [119]) suggest that sleep spindles are generated in the thalamus (a relay nucleus located near the center of the head) with propagation to the cortex via thalamocortical projections. This study was designed to test the hypothesis that the thalamus is the origin of sleep spindles in the human.

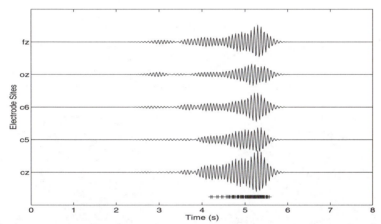

Fig. 11–4 Results of five channels of the reconstructed sleep spindle using time-frequency analysis and synthesis of the raw EEG shown in Fig. 11–1. A comparison between these two figures indicates that the background noise is effectively removed by the time-frequency technique. Dipoles are localized at the time slices indicated by "+" symbols below channel Cz.

We recorded spontaneous EEG (sampled at 256 Hz) from three sleep-deprived male subjects of ages 26, 28, and 39 using a 64-channel amplification and acquisition system. To avoid aliasing, an analog bandpass filter with cut-off frequencies of 0.1 Hz and 70 Hz were utilized before digitization. Electrodes were placed at the sites defined in the International 10-20 System and at the midpoints between these standard electrode sites. Sleep spindles were identified and notated by a trained sleep scorer. A total of 38 segments of EEG containing sleep spindels were identified and selected for this study. Each segment contained 64 channels of data with 2,048 samples per channel. In order to reduce the border effect in the process of computing time-frequency distributions, we positioned the spindle signal near the center of each segment. The segmented data were then down-sampled by a factor of four after digital lowpass filtering (cut-off frequency 25.6 Hz). A set of typical traces of the EEG (recorded from five of the electrode sites) has been previously shown in Fig. 11–1.

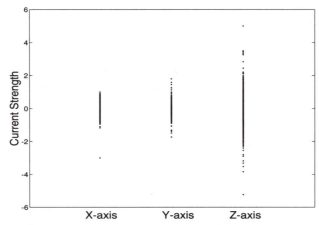

Fig. 11–5 Histogram of the strength of the dipole current in each of the x, y, and z coordinates. A more spread distribution of the current moment can be observed in the z-axis than in the x- and y-axes. This indicates that the source current of sleep spindles tends to be stronger in parallel to the direction of the vertex of the head.

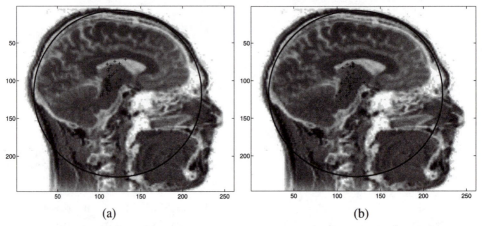

          (a)   (b)

Fig. 11–6 (a) Localized dipoles (with time-frequency processing) at "+" symbols in Fig. 11–6 are superimposed with the MRI sectional image of the subject. The dipoles (cross patterns) are concentrated in the neighborhood of the thalamus as suggested by the previous animal experiments. (b) Localized dipoles without time-frequency processing, but with the traditional bandpass filtering (Order-8 Butterworth) between 10 Hz and 16 Hz.

Finally, to compare these results with the ones obtained using traditional bandpass filtering (Order-8 Butterworth filter, cut-offs, 10 and 16 Hz), we localized dipoles at the same time slices corresponding to Fig. 11–6a. These results are plotted in Fig. 11–6b where the patterns are scattered due to the influence of noise. As shown in Table 11–1 , the average residual variance is 60% larger (0.0619 vs. 0.0389).

Table 11–1 Spindle Source Locations for Three Subjects

Subject	Locations	Stand Dev.	Res. Var.	No. Dipoles
No. 1	(0.1007,-0.0064,0.1961)	(0.1937,0.1129,0.1349)	6.09%	1,042
No. 2	(-0303,-0.0055,0.0948)	(0.1141,0.0729,0.1237)	3.35%	444
No. 3	(0649,0.0316,0.1999)	(0.1533,0.1064,0.1265)	5.43%	1,073

11.6 Conclusion

We have shown that time-frequency analysis and synthesis techniques are very useful for extracting signals from multichannel EEG. These techniques aid in exploring the inner workings of the brain by finding the sources that are responsible for the extracted EEG patterns. Our application of these techniques to the reconstruction of sleep spindles has provided a reliable input to the source localization algorithm.

Economic Data Analysis[*]

Because economic data, such as the stock price index, usually are good indicators of general economic trends and market sentiments, economic data analysis has attracted a great deal of attention in the area of signal processing. Historically, there are two approaches that dominate the financial community. One is developed by the fundamental school from academia and the other is advocated by the technical school from Wall Street.

Fundamentalists believe that the stock-price change follows a random walk. In other words, the history of prices has no power in predicting future prices. They also believe that economic movements always go toward equilibrium. Therefore, a proper trading strategy should be selling high and buying low, because the price disequilibrium cannot last long in an efficient market. A reasonable goal of buy-and-hold investment strategy, they suggest, is the average return. Fundamentalists ignore the existence of business cycles. Their theory is also weak in explaining market instability, such as the stock market crash.

People from the technical school gauge price levels and cyclic patterns from time series. Therefore, they are also known as *chartists*. Chartists make their trade decisions based on the trend and level information obtained from their graphics. They suggest that a rising trend signals the time to buy and a falling trend indicates the time to sell. A typical example of trading strategy is the so-called *five percentage rule*. That is, buy if the price goes up five percent or sell if

---

[*] Contributed by Ping Chen, Ilya Prigogine Center for Studies in Statistical Mechanics & Complex Systems, The University of Texas, Austin, Texas 78712

the price goes down five percent. The "level gauging" has a chance to beat the market, because price adjustments are usually slow and often over-reacted. However, the magnitude of unexpected shocks is not predictable. Consequently, it is hard to make the level gauging optimal. The trading strategy employed by the technical school is more like an art rather than a science.

Although it is impossible to precisely predict stock movements, we do see certain patterns possessed by the economic data series. For example, a stock index such as FSPCOM (which is also known as Standard & Poor 500 composite index, or simply, S&P 500), in Fig. 12–1, grows exponentially in general. Like FSPCOM, most aggregate economic indices can be roughly considered as a steady growth plus residual fluctuations. While the trend represents the long-term movement, fluctuations contain the information regarding short-term changes.

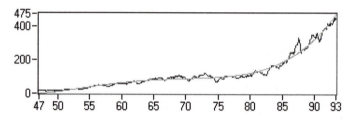

Fig. 12–1 FSPCOM (also known as S&P 500) series can be viewed as a smoothing trend plus a short period fluctuation. (Data source is CityBase.)

Traditionally, the fluctuation is treated as white noise. However, we have found that it is not true when the joint time-frequency analysis is applied. In fact, if the trend is *properly* removed, we could clearly observe some business cycles, though they evolve with time. These observations may fundamentally change current understanding of the market mechanism. It is the goal of this chapter to briefly introduce those exciting developments in the financial community.

As we know, for any given economic data, we could have many different ways to detrend them. Different detrending schemes will result in different fluctuations around the trend (the difference between the trend and raw data). Different fluctuations (or residuals, in econometric literature) will lead to different time-frequency patterns. So, the first issue for economic data analysis is how to detrend it so that the fluctuations lead to meaningful time-frequency patterns. In Section 12.1, we investigate, from a time-frequency analysis point of view, some well-established trend-cycle decomposition algorithms, such as log-linear detrending (LLD) and Hodrick-Prescott detrending (HP). Although joint time-frequency analysis remains a buzzword for most economists, it has been found to have great potential in the area of economics.

In Section 12.2, we apply the time-variant filter to extract deterministic cycles. In Section 12.3, we use time-frequency analysis to test historical events, such as the oil price shock in 1973 and the stock market crash in 1987. In Section

12.4, the potential of applications of time-frequency analysis to study deterministic color chaos is presented.

In this chapter, we mainly use monthly FSPCOM and FSDXP (S&P 500 dividend yield) to demonstrate the effectiveness of time-frequency analysis. FSPCOM consists of the 500 largest corporation stock prices, which is a most popular value-weighted index and has an important role in applied finance theory. Traditionally, FSPCOM is used as an indicator of general market trends. In the financial community, we also use the rate of returns of FSPCOM to evaluate investment performance.

12.1   Trend-Cycle Decomposition

The common techniques for economic data analysis is first to break up a complicated economic data series into two relatively simple subsets: long-term trend and short-period fluctuations (difference between the trend and raw data). We call such processing the *trend-cycle decomposition*. With the help of trend, we set the long-term framework. Based on the information extracted from the short-time fluctuations, we make a short term adjustment.

Obviously, the trend-cycle decomposition of a given data series is not unique. A different perception for a fluctuation pattern leads to a different detrending strategy. For example, fundamentalists believe that stock-price changes follow a random walk. They use the first difference (FD) of a logarithmic time series to find monthly growth ratio[*], i.e.,

$$\text{FD}(m) = \log s(m+1) - \log s(m) = \log \frac{s(m+1)}{s(m)} \tag{12.1}$$

where $s(m)$ denotes the raw data.

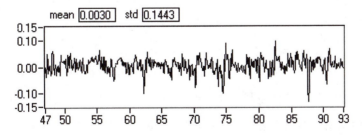

Fig. 12–2 FD for FSPCOM behaves as random noise (both mean and stand and deviation are computed based on log scale).

As shown in Fig. 12–2, the instantaneous growth rate in general is erratic. FD series appears as a typical random noise. Therefore, the process of FD is also

---

[*] Because most economic aggregate indexes have exponential growth trends, economists usually use a logarithm scale to linearize the curve.

called *pre-whitening* in econometric literature (that is, to make the fluctuation look like random noise).

Based on FD, the fundamentalist computes the mean and variance of stock-price changes. While the mean estimates an expected return, the standard deviation measures the risk of the investment. Applying FD, fundamentalists in fact idealize the economic movements. Based on their approach, the mean and variance of FD series are constant, or at least slow changing, which apparently is not what our everyday experience tells us.

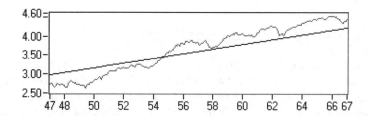

Fig. 12–3 LLD is linear, which implies that the growth rate is constant.

A more general detrending is the decomposition of the logarithmic time series into trend series and cycle series. The simplest one is the LLD, given by

$$\text{LLD}(m) = a + bm \qquad (12.2)$$

where coefficients a and b are computed by a regular linear regression algorithm based on the logarithmic time series $\log s(m)$. As shown in Fig. 12–3, LLD implies an idealized economic path with constant exponential growth.

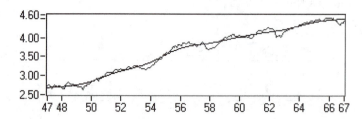

Fig. 12–4 Compared to LLD in Fig. 12–3, HP trend is closer to the raw data.

Fig. 12–4 illustrates the Hodrick-Prescott trend [81]. Hodrick and Prescott suggest that the desired trend should not only be smooth, but also has to be as close to the raw data as possible. Mathematically, the HP trend is determined by minimizing the following function

$$\sum [\log s(m) - \text{HP}(m)]^2 + \lambda \sum \{[\text{HP}(m+1) - \text{HP}(m)] - [\text{HP}(m) - \text{HP}(m-1)]\}^2$$

The first summation reflects the closeness between $\text{HP}(m)$ and the raw data. The second summation is the second difference of the trend, which is the measure of the smoothness of the trend. The parameter λ balances the closeness and smoothness. When $\lambda = 0$, $HP(m) = \log s(m)$. As λ gets larger, $HP(m)$ becomes smoother, but more different from $\log s(m)$. For $\lambda = \infty$, $HP(m)$ reduces to $LLD(m)$. In practice, $\lambda = 40^2$ for quarterly data and $\lambda = 120^2$ for monthly data [81].

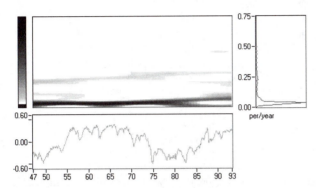

Fig. 12–5 The bottom plot is the time waveform of fluctuation $LLD_c(m)$. The long period of LLD cycles has no clear connection with business cycle theory.

Fig. 12–5 and Fig. 12–6 depict conventional power spectra and time-dependent spectra[*] for LLD cycles $LLD_c(m)$ and HP cycles $HP_c(m)$, respectively. The $LLD_c(m)$ and $HP_c(m)$ cycles are computed by

$$LLD_c(m) = \log s(m) - LLD(m) \tag{12.3}$$

and

$$HP_c(m) = \log s(m) - HP(m) \tag{12.4}$$

Although the LLD trend (see Fig. 12–3) has a smoother trend than HP's, LLD cycle $LLD_c(m)$ does not possess any useful patterns in business cycle theory. For government policy making, the monthly data are good enough to study business cycles that are in the range of two to seven years. Most reliable economic data are shorter than fifty years, which are not long enough to judge the cycle pattern longer than twenty years.

The characteristic period of $HP_c(m)$ is pretty close to our perception of business cycles. The time-dependent spectrum in Fig. 12–6 exhibits a stable four-year business cycle (0.25 period/year), one U.S. presidential term. This observation confirms a consensus among most economists, that is, economical movements are strongly influenced by institutional cycles in politics. In contrast, no relevant economical business cycles appear in the LLD case.

Although the conventional power spectra (right plots in Fig. 12–5 and Fig. 12–6) also indicate the four-year cycle, it is not clear, based on the power spectra

---

[*] All time-dependent spectra presented in this chapter are computed by the low-order time-frequency distribution series introduced in Chapter 7.

alone, whether or not the four-year cycle has dominated the last fifty years. On the other hand, the time-dependent spectra well delineate how business cycles change over time, which enables us to better understand the economic mechanism.

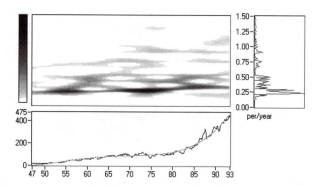

Fig. 12–6 The bottom plot is the time waveform of fluctuation $HP_c(m)$. The time-dependent spectra clearly exhibits a stable four-year business cycle (0.25 period/year).

It has been found that the detrending scheme has substantial effects on the results of joint time-frequency analysis. The less smooth the trend is, the shorter the implied period of cyclic fluctuations. For government policy studies, HP is desirable. However, for speculative investors, high-frequency samples (such as days or minutes) are needed. For their applications, the applicable detrending algorithm should reflect shorter cycles than that obtained by the HP method. This is an ongoing research topic. In this section, we only discuss cases that most economists find interesting.

We have shown that, by using joint time-frequency analysis, we could obtain some interesting information that is not available in either time or frequency representation alone. In what follows, we shall briefly discuss the significance of those new findings for the economic community.

12.2 Extraction of Characteristic Period via Time-Variant Filter

By applying joint time-frequency analysis, we see that most economic movements contain strong time-varying cycles as well as substantial random noise. Because observed deterministic cycles evolve over time, it is difficult to effectively separate them from random noise by conventional techniques, such as a bandpass filter.

It is interesting to note that while the noise tends to evenly spread into the entire joint time-frequency domain, deterministic cycles are continuous and concentrated in the time-frequency domain. Consequently, we can apply the time-variant filter, introduced in Chapter 9, to extract time-varying deterministic cycles.

Fig. 12–7 illustrates the original time series as well as the filtered economic time series. Obviously, the filtered economic time series is much smoother. Moreover, it closely resembles the original time series. The correlation coefficient between the filtered and original series is 0.85. Their ratio of standard deviation is 85.8%. This suggests that the variation of the stock price mainly follows relatively simple patterns. Random noise only plays a minor role in business cycle movements. Such an observation is close to the chartist's intuition and contradicts the fundamentalist's assertion.

The magnitude of random noise is a critical parameter in finance theory. So far we do not know the pricing mechanism of stock prices. But we do know how to estimate the price of derivative securities whose value depends on the associated security price. A proper measurement of standard deviation is critical in option pricing theory [12]. Economists are puzzled by the gap between implied volatility derived from option prices and historical volatility of stock prices calculated from FD detrending. A consistent volatility may be obtained by a proper detrending.

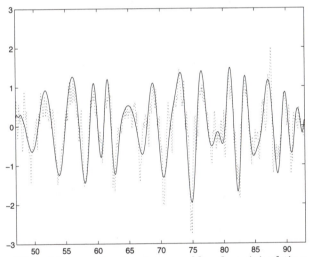

Fig. 12–7 The filtered time series closely resembles the original time series. The correlation coefficient between the filtered and original series is 0.85.

From the time-frequency representation of a filtered time series, we can define a characteristic period by tracing the dominant instantaneous frequencies. For the deterministic process, the time path is a continuous line in the time-frequency domain. From our analysis of a large number of economic indicators, we have found that economic indicators often have their distinctive patterns in time-frequency representation. The characteristic period time pattern can be considered as the "fingerprint" of an underlying deterministic process. It can be used for economic diagnosis.

One important finding is that stock market indicators have quite similar patterns from their characteristic periods, such as those in FSPCOM and FSDXP in Fig. 12–8. It is not accidental that FSPCOM and FSDXP demonstrate similar patterns during most historical shocks. This observation suggests that the economic system acts in fact like an organism. Its time rhythm can be an indicator of adapting to environmental changes.

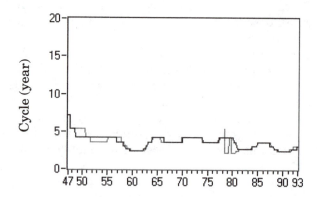

Fig. 12–8 Characteristic periods for FSPCOM and FSDXP yields.

Chartists often use level knowledge in addition to trend knowledge in speculative trading. Because the magnitude of unexpected shocks is not predictable, the "level gauging" method is not reliable in forecasting turning points. However, the characteristic frequency of the stock market is more stable over time. This observation can lead to a new strategy of "period trading": buy in a rising period and sell before a falling period.

12.3 Economic Diagnosis of Historical Events

Like all other sciences, one primary interest of economists has been the relationship between events and their causes. For example, what causes market crises? Are they caused by an external reason or an economic system's internal instability? If we consider the economic system as an organism rather than a sandpile, then business cycles can be thought of as a heart rhythm that provides some clues to diagnose the historical economic events.

Two dramatic events in recent economic history were the oil price shock and stock market crash. In October 1973, OPEC (Organization of Petroleum-Exporting Countries) raised the price of oil 66% and doubled it three months later in January 1974. This action shocked the stock market. That stock-price disturbance, as most economists now agree, apparently was external in nature. This assertion can also be verified by a joint time-frequency analysis.

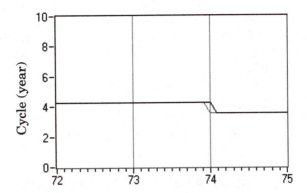

Fig. 12–9 Characteristic periods of FSDXP and FSPCOM were both shifted after the oil price shock in October 1973. It is suggested that corresponding stock market changes were caused by external forces, the oil price shock.

Fig. 12–9 depicts characteristic periods of FSDXP and FSPCOM between 1972 to 1975. Note that both of them were shifted after OPEC raised the price of oil, which indicates that the stock-price change was caused by an external force.

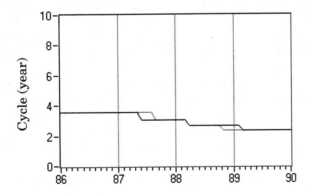

Fig. 12–10 Characteristic periods of both FSPCOM and FSDXP changed before the stock market crash in October 1987. It is suggested that the stock market crash resulted from some internal instabilities.

While the nature of the oil price shock is relatively easy to understand, there is less agreement about the cause of the stock market crash on October 19, 1987, when stock prices plunged more than 25% within one day. For those who suspect external causes, there were not any scapegoats. For those who believe in internal reasons, there was no solid evidence. Fig. 12–10 shows that as early as four months before the stock market crash, characteristic periods of both FSP-COM and FSDXP already moved. If we accept that the cause comes before the effect, then we would conclude that the stock market crash was caused by an internal instability of the economic system. It is worth noting that such observation cannot be obtained by any currently used analysis techniques.

Moreover, the extraordinary resilience of the stock market can be revealed from the variability of the characteristic period under shocks. These events generated only minor changes, 17% in 1974 and 14% in 1987, in characteristic periods of FSPCOM and FSDXP.

As our testing indicated, the characteristic period path can be a useful tool in economic diagnosis. Further studies of time-frequency patterns of pertinent economic indicators may shed new light on historical events, such as military spending and changing monetary, fiscal, or tax policies.

12.4   Time-Frequency Analysis for Detecting Deterministic Chaos

The recent development of non-linear dynamics suggests that business cycles may be better understood and modeled by the concept of *color chaos*. Color chaos here refers to the non-linear oscillator generated by a low-dimensional continuous-time non-linear deterministic system, which can be characterized by non-linear differential equations or non-linear difference-differential equations. Linear oscillators are characterized by an integer dimensionality. For example, the dimension of a periodic motion is one. The dimensionality of deterministic chaos is a fractal number (see [18] and [65]). Conceptually, we can consider that chaos is a mathematical model that behaves between the two extreme models, linear oscillator and white noise. In other words, it is not as simple as the linear oscillator, but it is also not completely random.

The main tasks of applying chaos theory are: detecting of deterministic chaos and characterizing the chaos system. Without delving into any details, we should point out that there is no single approach that is able to exclusively identify chaos. Usually, we have to employ a package of complementary approaches to reveal the existence of deterministic chaos [19]. In most cases, however, the continuous-time color chaos possesses well-defined deterministic cycles.

The difficulty of quantitatively describing chaos systems mainly is due to the pollution of random noise. In early studies of experimental chaos, spectral analysis played an important role [57]. Because deterministic cycles in non-controllable experiments usually evolve over time, it is hard to effectively separate them from background noise by the traditional bandpass filter. In this section, we shall see that joint time-frequency analysis would be a good alternative for those conventional techniques.

The Fourier spectrum of white noise is flat. On the other hand, the linear harmonic oscillator appears as sharp peaks at the basic frequency and its higher harmonics, such as $2f$, $3f$, ... The "color" of chaos implies that a characteristic frequency exists over a noisy background. The time series of color chaos looks like random noise, whereas its spectrum has a wider cluster and sub-harmonic frequencies at $f/2$, $f/3$, ... Based on our previous discussions, it is obvious that the time-dependent spectrum is much more powerful in detecting deterministic cycles than either time or frequency representations alone.

From the time-dependent spectrum, we first can check whether time-varying deterministic cycles exist. If so, it would be important evidence of continuous-time deterministic chaos. Then, we can apply the time-variant filter to effectively separate background noise and useful time-varying patterns. It is much easier to describe the clean deterministic cycle mathematically than noise-corrupted signals.

A popular tool in non-linear dynamics is the phase portrait in the phase space. The phase portrait is a 2D plot, $X(m+T)$ vs. $X(m)$, where T is a fixed time lag in the order of characteristic period. Obviously, the phase portrait reflects the structural relationship between $X(m + T)$ and $X(m)$. Because white noise is uncorrelated, its phase portrait appears structureless. On the other hand, deterministic cycles usually possess clear patterns.

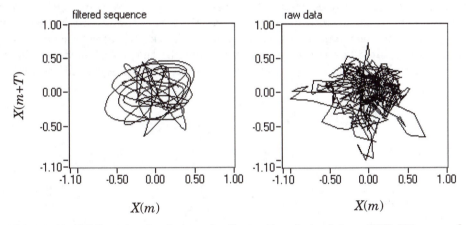

Fig. 12–11 While noise dominates the fluctuation derived from S&P 500, a much clearer pattern appears in a filtered sequence. The time delay T = 60 months.

Fig. 12–11 compares the filtered FSPCOM HP cycles with the original FSCOM HP cycles. The phase portrait of filtered FSPCOM HP cycles shows a clear pattern of deterministic spirals, a typical feature of color chaos. Standard tests in non-linear dynamics show more evidence of deterministic chaos from filtered time series. Its fractal dimension is about 2.5 (see [19] and [20]).

The finding of color chaos from the stock index may build a bridge between the fundamental school and the technical school in finance theory, which would explain the difficulty in economic forecasting encountered by fundamentalists. Chaotic movements have an intrinsic limit of predictability. In addition to risk characterized by normal distribution of stock-price changes, investors are facing additional uncertainties: the changing market trends and changing characteristic frequency. Although business cycles have significant components of deterministic chaotic cycles, we only have a limited predictability. Sophisticated investors armed with time-frequency analysis may have a better chance to beat average investors, but no one can be a sure winner in an evolving economy.

Fundamentalists may argue that perfect competition will eventually eliminate deterministic patterns in the stock market, if people follow the winner will drive down any profit margin. This possibility can be true only in the idealized linear world with perfect information and instant adjustment. In practice, nonlinear interaction and time-delay in response often increase overshooting of business cycles. That is why market fluctuations always coexist with business cycles.

Conclusion

In this section, we demonstrated that joint time-frequency analysis has great potential in studying stock market movements. By properly detrending, the complicated economic behavior could be characterized by a long-term trend plus short-time business cycles. Using the time-variant filter, we can largely remove the random noise in stock-price changing. Moreover, time-frequency analysis may be used for economic diagnosis of historical shocks and economic forecasting of turning points. The characteristic frequency, or the "color" of deterministic cycles, provides further clues of color chaos, a hot topic in advanced research in physics, as well as economics.

Appendices

A. Critical Sampling Discrete Gabor Expansions

This appendix is devoted to the proof of the necessary and sufficient condition of $h[k]$ of the discrete Gabor expansion at critical sampling[*]. For critical sampling, $\Delta M = N$, (3.49) reduces to

$$\sum_{k=0}^{L-1} \bar{h}[k + qN] W_N^{-pk} \gamma^*[k] = \delta[p]\delta[q] \tag{A.1}$$

for $0 \leq p < N$ and $0 \leq q < 2(\Delta N - 1)$. Then (A.1) can be written in a matrix form as

$$\bar{H}\bar{\gamma}^* = \begin{vmatrix} W & 0 & 0 & 0 & 0 & & 0 \\ 0 & W & 0 & 0 & 0 & \ldots & 0 \\ 0 & 0 & W & 0 & 0 & & 0 \\ 0 & 0 & 0 & W & 0 & \ldots & 0 \\ 0 & 0 & 0 & 0 & W & & 0 \\ & \ldots & & \ldots & & \ldots & \\ 0 & 0 & 0 & 0 & 0 & & W \end{vmatrix} \begin{vmatrix} H_0 & H_1 & \ldots & H_{\Delta N-1} \\ H_1 & \ldots & H_{\Delta N-1} & 0 \\ \ldots & H_{\Delta N-1} & 0 & \ldots \\ H_{\Delta N-1} & 0 & \ldots & 0 \\ 0 & \ldots & 0 & H_0 \\ \ldots & & \ldots & \\ 0 & H_0 & \ldots & H_{\Delta N-2} \end{vmatrix} \bar{\gamma}^* = \bar{\mu}$$

$$\tag{A.2}$$

where $\bar{H}$ is $(2\Delta N\text{-}1)$N-by-L block matrix. Each block H_q is an N-by-N diagonal matrix given by

$$H_q = \begin{vmatrix} h[qN] & 0 & \ldots & 0 \\ 0 & h[qN+1] & \ldots & 0 \\ \ldots & & \ldots & \\ 0 & & \ldots & 0 & h[qN+(N-1)] \end{vmatrix}$$

W is an N-by-N DFT matrix, a special *Hermitian* matrix, i.e.,

$$W = \begin{vmatrix} 1 & 1 & 1 & \ldots & 1 \\ 1 & W^{-1} & W^{-2} & \ldots & W^{-(N-1)} \\ 1 & W^{-2} & W^{-4} & \ldots & W^{-2(N-1)} \\ \ldots & & .. & & \ldots \\ 1 & W^{-(N-1)} & W^{-2(N-1)} & \ldots & W^{-(N-1)(N-1)} \end{vmatrix}$$

where $W = \exp(j2\pi/N)$. Because $W^{-1} = W^*/N$, multiplying both sides of Eq. (A.2)

---

[*] Proved by Zhanbo Yang, Department of Mathematics, Shawnee State University, Ohio 45662.

by W^{-1} yields

$$
\begin{vmatrix}
H_0 & H_1 & \cdots & H_{\Delta N-1} \\
H_1 & \cdots & H_{\Delta N-1} & 0 \\
\cdots & H_{\Delta N-1} & 0 & \cdots \\
H_{\Delta N-1} & 0 & \cdots & 0 \\
0 & \cdots & 0 & H_0 \\
\cdots & & \cdots & \\
0 & H_0 & \cdots & H_{\Delta N-2}
\end{vmatrix}
\vec{\gamma}* =
\begin{vmatrix}
\vec{u} \\
0 \\
0 \\
0 \\
0 \\
\cdots \\
0
\end{vmatrix}
\qquad \text{(A.3)}
$$

where

$$
\vec{u} = (\underbrace{\frac{1}{N}, \frac{1}{N}, \dots \frac{1}{N}}_{N})^T
\qquad \text{(A.4)}
$$

For presentation convenience, we write (A.3) as

$$
H\vec{\gamma} = \vec{v}
\qquad \text{(A.5)}
$$

where H is a block matrix as indicated in (A.3). In each block, there are N rows.

T H E O R E M

(1) *The linear system* (A.1) *has a solution* $\gamma[k]$ *iff for each* k, $k = 0,1,\dots N-1$, *the set* $\{h[qN+k] : q = 0,1,\dots,\Delta N-1\}$ *is exclusively non-zero, i.e.,*

$h[k] \otimes h[N+k] \otimes h[2N+k] \otimes \dots h[(\Delta N-1)N+k] \neq 0.$

(2) *If* (A.1) *has a solution, then it is unique and given by*

$$
\gamma[j] = \begin{cases}
\dfrac{1}{Nh[j]} & h[j] \neq 0 \\[2ex]
0 & \text{otherwise}
\end{cases}
$$

where $j = 0, 1, 2,\dots, L-1$.

PROOF

Since (A.5) is equivalent to (A.1), we will discuss the proof with respect to (A.5) only.

Necessity $(\Rightarrow)$

From a standard linear algebra theorem, a linear system (A.5) has a solution *iff* the coefficient matrix H and the augmented matrix $|H\vec{v}|$ have the same rank. In other words, the solution of (A.5) exists *iff* the last column of the augmented matrix, namely $\vec{v}$, is a linear combination of the first L many columns, namely the columns of H. This implies that there are L constants, $a_j, j = 0, 1, 2,..., L{-}1$, not all zero, such that

$$\sum_{j=0}^{L-1} a_j \vec{h}_j = \vec{v} \tag{A.6}$$

Recall that H can be partitioned into $(2\Delta N{-}1)$ row blocks. And each block consists of N rows. Then (A.6) can be written in terms of $(2\Delta N{-}1)N$ linear equations. For example,

At the *zeroth* row block, the kth linear equation is

$$\sum_{q=0}^{\Delta N-1} a_{qN+k} h[qN+k] = \frac{1}{N} \tag{A.7}$$

At the *nth* row block, where $n = 1,2,...,\Delta N{-}1$, the kth linear equation is

$$\sum_{q=0}^{\Delta N-1-n} a_{qN+k} h[(q+n)N+k] = 0 \tag{A.8}$$

At the *nth* row block, where $n = \Delta N, \Delta N{+}1,...,2\Delta N{-}2$ the kth linear equation is

$$\sum_{q=0}^{n-\Delta N} a_{(\Delta N-1-q)N+k} h[(q+n-\Delta N)N+k] = 0 \tag{A.9}$$

In (A.7) to (A.9), $k = 0, 1,..., N{-}1$. Now, let's prove the following proposition with the mathematical induction[*].

Proposition:

For each $p = 0,1,...,\Delta N{-}1$,

(a) $a_{pN+k} \neq 0$ implies that $h[pN+k] \neq 0$ and all $h[mN+k] = a_{mN+k} = 0$ for all $m = 0, ... , \Delta N{-}1$ and $m \neq p$

(b) $h[pN+k] \neq 0$ implies that

$$a_{pN+k} = \frac{1}{Nh[pN+k]} \neq 0$$

and all $h[mN+k] = a_{mN+k} = 0$ for all $m = 0,1,...,\Delta N{-}1$ and $m \neq p$.

---

[*] Mathematical induction principle: Let $P(n)$ be a proposition which depends on an arbitrary natural number n, if (1) $P(0)$ is true, (2) the assumption that $P(0),..., P(n{-}1)$ are true implies that $P(n)$ is true, then the proposition is true for all natural numbers n.

Proof of the Proposition:

First, let $p = 0$.

(a) if $a_k \neq 0$, then for $n = \Delta N - 1$ (from (A.8)) we have

$$a_k \, h[(\Delta N - 1)N + k] = 0$$

which implies $h[(\Delta N - 1)N + k] = 0$. Using this result for $n = \Delta N - 2$, we obtain

$$a_k \, h[(\Delta N - 2)N + k] + a_{N+k} \, h[(\Delta N - 1)N + k] = 0$$

yields

$$a_k \, h[(\Delta N - 2)N + k] = 0$$

which implies

$$h[(\Delta N - 2)N + k] = 0$$

Continuing this process up to $n = 1$ yields

$$h[mN + k] = 0 \qquad\qquad\qquad\qquad (A.10)$$

for $1 \leq m < \Delta N$. Applying (A.10) into (A.7) leads to

$$a_k h[k] = \frac{1}{N}$$

which implies $h[k] = 1/(a_k N) \neq 0$. Now, let's examine $n = \Delta N$ based on (A.9).

$$a_{(\Delta N - 1)N + k} \, h[k] = 0$$

Because $h[k] \neq 0$, the above equation implies $a_{(\Delta N - 1)N + k} = 0$. Using this result for $n = \Delta N + 1$, we obtain $a_{(\Delta N - 2)N + k} = 0$. Continuing this process up to $n = 2\Delta N - 2$ yields

$$a_{mN + k} = 0 \qquad\qquad\qquad\qquad (A.11)$$

for $1 \leq m < \Delta N$.

End of proof for part (a) at $p = 0$

(b) Because (A.7) to (A.9) are symmetric in $h[mN+k]$ and a_{mN+k}, a similar process can be used to show that if $h[k] \neq 0$, then all other $h[mN+k]$ and a_{mN+k} are zero for $m = 1,...,\Delta N - 1$.

End of proof for the part (b) at $p = 0$

Now, let's assume that the proposition is true for $p = 0,...,q-1 < \Delta N - 1$. We are going to show that it is also true for the case of $p = q$.

Since the induction hypotheses assume that the proposition holds for $p = 0,...,q-1 < \Delta N - 1$, it must be true that if (a) $a_{pN+k} \neq 0$ or (b) $h[pN+k] \neq 0$, then $a_{mN+k} = h[mN+k] = 0$ for $m = 0,...,q-1, q, ... \Delta N-1$ and $m \neq p$. In other words, if (a) $a_{pN+k} \neq 0$ or (b) $h[pN+k] \neq 0$, then $a_{pN+k} = h[pN+k] = 0$ for $p = 0,...,q-1 <$

$\Delta N-1$ (otherwise, it will contradict the induction hypotheses). The remaining question is what about a_{mN+k} and $h[mN+k]$ for $m = q, q+1,...,\Delta N-1$ when (a) $a_{qN+k} \neq 0$ or (b) $h[qN+k] \neq 0$. In the following, we will prove that if (a) $a_{qN+k} \neq 0$ or (b) $h[qN+k] \neq 0$, then $a_{mN+k} = h[mN+k] = 0$ not only for $m = 0,1,2,...,q-1$ (as shown above) but also for $m = q, q+1, ..., \Delta N-1$.

Let's rewrite (A.7) to (A.9) by omitting all $0 = 0$ (caused by $a_{pN+k} = h[pN+k] = 0$ for $p = 0,...,q-1 < \Delta N-1$) as

At $n = 0$, the kth linear equation is

$$\sum_{m=q}^{\Delta N-1} a_{mN+k} h[mN+k] = \frac{1}{N} \tag{A.12}$$

At $n = 1,2,..., \Delta N-1-q$, the kth linear equation is

$$\sum_{m=q}^{\Delta N-1-n} a_{mN+k} h[(m+n)N+k] = 0 \tag{A.13}$$

At $n = (2\Delta N-1)-q,...,2\Delta N-2$, the kth linear equation is

$$\sum_{m=q}^{n-\Delta N} a_{(\Delta N-(1-m))N+k} h[(m+n-\Delta N)N+k] = 0 \tag{A.14}$$

The rest of the proof is similar to the case $p = 0$. For example, starting with $n = (\Delta N-1)-q$ from (A.13) with $a_{qN+k} \neq 0$, one can prove that $h[qN+1] \neq 0$ and $a_{mN+k} = h[mN+k] = 0$ for $m \neq q$. Then, examine $n = (2\Delta N-1)-q$ from (A.14) with $h[qN+1] \neq 0$. Obtain $a_{mN+k} = h[mN+k] = 0$ for $m \neq q$.

End of proof for parts (a) and (b) at $p = q$

By the induction principle, the proposition is true for all $p = 0,...,k-1$.

End of proof for the proposition

This proposition implies that if (A.5) has a solution, then each set $\{h[mN+k] : m = 0,1,...,\Delta N-1\}$ must be *exclusively non-zero*. Moreover,

$$a_{mN+k} = \begin{cases} \dfrac{1}{Nh[mN+k]} & h[mN+k] \neq 0 \\ 0 & \text{otherwise} \end{cases}$$

End of proof for the necessity

Sufficiency ($\Leftarrow$)

Assume that each set $\{h[mN+k] : m = 0,1,...,\Delta N-1\}$ is exclusively non-zero for $k = 0,..., N-1$. Let $h[mkN+k] \neq 0$ and

$$a_{m_kN+k} = \frac{1}{Nh[m_kN+k]}$$

and $h[m_jN+j] = a_{mjN+j} = 0$ *for* $j \neq i$. It is easy to see that equation set (A.7) to (A.9) are satisfied. In other words, Eq. (A.5) has a solution.

End of proof for part (1) of the theorem

(2) If (A.5) does have a solution, then by part (1) of this theorem, for each index $k = 0,...,N-1$, the set $\{h[mN+k]) : m = 0,1,...,\Delta N-1\}$ is *exclusively non-zero*. There is an index m_k such that $h[m_kN+k] \neq 0$. It is easy to verify that

$$\gamma[mN+k] = \begin{cases} \dfrac{1}{Nh[mN+k]} & h[mN+k] \neq 0 \\ 0 & \text{otherwise} \end{cases}$$

is the solution.

End of proof

B. Optimal Dual Functions

For a given function $h[k]$, the corresponding dual function $\gamma[k]$ may not be unique. In Section 3.5, we discussed the solution of $\gamma[k]$ that is optimally close to $h[k]$. The resulting Gabor expansion is called the orthogonal-like Gabor expansion.

From the definition, it is clear that the Gabor transform and Gabor expansion can be considered as perfect reconstruction filter banks. In those applications, the analysis and synthesis functions could often have different requirements; one may intentionally pursue $\gamma[k]$ that differs from $h[k]^*$. In this section, we shall investigate a general optimal algorithm that allows $\gamma[k]$ to be optimally close to an arbitrarily desired function $d[k]$, for a given $h[k]$.

We formulate the problem as follows. For a given function $h[k]$ and the sampling pattern (determined by the time sampling interval ΔM and the number of frequency channels N), find a dual function $\gamma[k]$ that is most similar to, in the sense of least square error (LSE), a desired function $d[k]$, i.e.,

$$\Gamma = \min_{\gamma : H\gamma^* = \vec{\mu}} \left\| \frac{\vec{\gamma}}{\|\vec{\gamma}\|} - \vec{d} \right\|^2 \tag{A.15}$$

where $d[k]$ is a unit energy function. Obviously, when $d[k] = h[k]$, (A.15) reduces to (3.56), which leads to the orthogonal-like Gabor expansion.

In general, (A.15) is a least square problem with an equality constraint given by

$$H\vec{\gamma^*} = \vec{\mu} \tag{A.16}$$

where H is a p-by-L matrix and $\vec{\mu}$ is a vector with p-element. Both of them are defined in Section 3.5. Without loss of generality, let's assume that H has a full row rank. (Otherwise, we always can employ SVD to alleviate the rank deficiency problem as introduced in Section 3.5.)

By QR decomposition [58], we have

$$H_{LP}^T = Q_{LL} \begin{bmatrix} R_{pp} \\ 0 \end{bmatrix} \tag{A.17}$$

Where Q is orthonormal and R is upper triangular. Substituting (A.17) into (A.16) obtains

$$\begin{bmatrix} R^T & 0 \end{bmatrix} Q^T \vec{\gamma} = \begin{bmatrix} R^T & 0 \end{bmatrix} \begin{bmatrix} \vec{x} \\ \vec{y} \end{bmatrix} = \vec{\mu} \tag{A.18}$$

---

[*] As mentioned in previous sections, $h[k]$ and $\gamma[k]$ are dual functions; they are exchangeable. We can use $\gamma[k]$ as analysis filter and $h[k]$ as synthesis filter, or vise versa.

where

$$Q^T \vec{\gamma} = \begin{bmatrix} \vec{x} \\ \vec{y} \end{bmatrix}$$

(A.19)

From (A.18),

$$\vec{x} = (R^T)^{-1}\vec{\mu}$$

(A.20)

Because $QQ^T = I$, left multiplying Q to both sides (A.19) yields

$$\vec{\gamma} = Q\begin{bmatrix} \vec{x} \\ \vec{y} \end{bmatrix} = \begin{bmatrix} Q_x & Q_y \end{bmatrix}\begin{bmatrix} \vec{x} \\ \vec{y} \end{bmatrix} = Q_x\vec{x} + Q_y\vec{y}$$

(A.21)

where

$$x \in R^P \qquad y \in R^{L-p}$$

Hence, $\vec{\gamma}$ is the sum of two orthogonal vectors,

$$Q_x\vec{x} + Q_y\vec{y}$$

Consequently,

$$\left\|\vec{\gamma}\right\|^2 = \left\|\vec{x}\right\|^2 + \left\|\vec{y}\right\|^2$$

(A.22)

Expanding the error formula (A.15) yields

$$\Gamma = \min_{\gamma:H\gamma^* = \vec{\mu}} \left\| \frac{\vec{\gamma}}{\left\|\vec{\gamma}\right\|} - \vec{d} \right\|^2 = \min_{\gamma:H\gamma^* = \vec{\mu}} 2\left(1 - \frac{Re(\vec{\gamma}^T\vec{d})}{\left\|\vec{\gamma}\right\|} \right)$$

(A.23)

Then, minimizing Γ with respect to $\vec{\gamma}$ is equivalent to

$$\max_{\gamma:H\gamma^* = \vec{\mu}} \frac{Re(\vec{\gamma}^T\vec{d})}{\left\|\vec{\gamma}\right\|} = \max_{\gamma:H\gamma^* = \vec{\mu}} \xi$$

(A.24)

Replacing $\vec{\gamma}$ with (A.21) and $\left\|\vec{\gamma}\right\|$ with (A.22), Eq. (A.24) becomes

$$\max_{y \in R^{L-p}}\xi = \max_{y \in R^{L-p}} \frac{Re(\vec{x}^T Q_x^T \vec{d}) + Re(\vec{y}^T Q_y^T \vec{d})}{\sqrt{\left\|\vec{x}\right\|^2 + \left\|\vec{y}\right\|^2}}$$

(A.25)

It is interesting to note that the maximum of ξ occurs when $\vec{y}$ is in the same direction as the vector $Q_y^T \vec{d}$, regardless of the value of $\left\|\vec{y}\right\|$. Let

$$\vec{y} = tQ_y^T \vec{d}$$

(A.26)

where t is real and positive. Then the problem of (A.25) is equivalent to maximizing ξ with respect to t. By replacing $\vec{y}$ with (A.26), (A.25) becomes

$$\max_{t>0} \frac{Re(\vec{x}^T Q_x^T \vec{d}) + \left\|Q_y^T \vec{d}\right\|^2 t}{\sqrt{\|\vec{x}\|^2 + \left\|Q_y^T \vec{d}\right\|^2 t^2}} = \max_{t>0} \frac{A + Bt}{\sqrt{C + Bt^2}} = \max_{t>0} \xi \qquad (A.27)$$

where

$$A = Re(\vec{x}^T Q_x^T \vec{d}) \qquad B = \left\|Q_y^T \vec{d}\right\|^2 \qquad C = \|\vec{x}\|^2$$

Obviously,

$$\lim_{t \to \infty} \xi = \sqrt{B} = \left\|Q_y^T \vec{d}\right\| \qquad (A.28)$$

For $A = 0$,

$$\xi' = \frac{BC}{\sqrt{(C + Bt^2)^3}} \qquad (A.29)$$

In this case, ξ does not have an extreme for the finite t. In other words, (A.15) does not have a solution if the desired function $\vec{d}$ is perpendicular to $Q_x \vec{x}$.

When $A \neq 0$, the first derivative of ξ, with respect to t, equal to zero yields an unique solution $t = C/A$. Next, we compute the second derivative to see whether or not $t = C/A$ corresponds to the maximum of ξ.

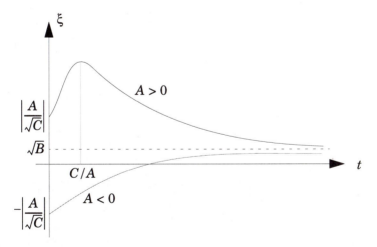

Fig. A–1 Error curves (ξ attains the maximum only for $A > 0$).

The second derivative of ξ with respect to t is

$$\xi'' = \frac{BC}{\sqrt{(C + Bt^2)^5}} (2ABt^2 - 3BC - AC) \qquad (A.30)$$

Replacing $t = C/A$, (A.30) becomes

$$\xi'' = -\frac{C}{A}(BC + A^2)$$
(A.31)

When $A > 0$, $\xi(t)$ is convex and attains its maximum at $t = C/A > 0$. When $A < 0$, $\xi(t)$ is concave and increases to $\left\| Q_y^T \vec{\partial} \right\|$ as t goes to infinity; this corresponds to an unbounded $\vec{y}$, which is impractical. Hence, the solution of (A.15) exists only for $A = Re(\vec{x}^T Q_x^T \vec{\partial}) > 0$. Fig. A–1 depicts the error curves.

Substituting (A.26) and $t = C/A$ into (A.21), we can readily obtain the solution of (A.15) as

$$\vec{\gamma} = Q_x \vec{x} + \frac{\|\vec{x}\|^2}{Re(\vec{x}^T Q_x^T \vec{\partial})} Q_y Q_y^T \vec{\partial}$$
(A.32)

and the minimum error is

$$\Gamma = 2\left(1 - \sqrt{\frac{\left\| Re(\vec{x}^T Q_x^T \vec{\partial}) \right\|^2}{\|\vec{x}\|^2} + \left\| Q_y^T \vec{\partial} \right\|^2}\right)$$
(A.33)

C. Existence of Adaptive Representations

The convergent properties of the adaptive representation, introduced in Section 8.2, are valid only under limited cases where finite signal points are used. In those circumstances, the rate of convergence is indeed exponential. The residuals eventually vanish. In what follows, we shall investigate more general cases. As a matter of fact, the adaptive matching process is always convergent, but the residuals are not necessarily equal to zero[*].

Let's define S as a *Hilbert* space and V as any closed subspace in S constructed by an orthogonal projection P such that $P: X \rightarrow V$. It is a linear map completely characterized by the property

$$s - Ps \perp V, (s \in S) \tag{A.34}$$

Eq.(A.34) indicates that the mapping function P divides S into two subspecies, V and an orthogonal complement of V. Assuming for the moment that such a map exists, we can show that Ps is the unique point in V closest to s. To do so, let v be any point in V; due to property (A.34) we have

$$\|s - v\|^2 = \|(s - Ps) + (Ps - v)\|^2 = \|s - Ps\|^2 + \|Ps - v\|^2 \geq \|s - Ps\|^2 \tag{A.35}$$

This shows that the distance between a point s in S and any point in V is always greater than or equal to the distance between s and its orthogonal projection point Ps. The existence of Ps as the closest point to s comes from the basis theorem that for any closed convex set in a Hilbert space, each point of the space has a unique closest point in the convex set. To prove that Ps, as the closest point to s, satisfies , let $v \in V$, $v_0 = Ps$ and $\lambda > 0$. Because v_0 is the closest point to s, we have

$$0 \leq \|s - v_0 - \lambda v\|^2 - \|s - v_0\|^2 = 2\langle s - v_0, \lambda v \rangle + \lambda^2 \|v\|^2 \tag{A.36}$$

Since $\lambda > 0$, we have

$$0 \leq 2\langle s - v_0, v \rangle + \lambda \|v\|^2 \tag{A.37}$$

Letting $\lambda \downarrow 0$, we obtain $<s-v_0,v> \geq 0$. Since v is an arbitrary point in V, for the point $-v$ in V we also have $<s-v_0,-v> \geq 0$ which leads to $<s-v_0,v> \leq 0$. Therefore, we must conclude $<s-v_0,v> = 0$. Hence, $(s-v_0) \perp v$ and $(s-v_0) \perp V$. v_0 is the projection point of s and $s-v$ is the residual part, as discussed in the previous sections. To also prove that P is linear, we note that $(\alpha x + \beta y) - (\alpha Px + \beta Py) \perp V$ because for $v \in V$,

$$\langle \alpha x + \beta y - \alpha Px - \beta Py, v \rangle = \alpha \langle x - Px, v \rangle + \beta \langle y - Py, v \rangle = 0 \tag{A.38}$$

Therefore $(\alpha Px + \beta Py) = P(\alpha x + \beta y)$, by the unicity of the closest point.

---

[*] Proved by Junjiang Lei and E.Ward Cheney, Dept. of Mathematics, University of Texas, Austin, Texas 78712-1084.

   The easiest theoretical construction of mapping P is to use any orthonormal base $\{u_\alpha\}$ for V. Then

$$Ps = \sum \langle s, u_\alpha \rangle u_\alpha \tag{A.39}$$

It is easy to define that the mapping P defined in Eq.(A.39) has the property shown in (A.34), because for each u_β we have

$$\langle s - Ps, u_\beta \rangle = \langle s, u_\beta \rangle - \langle \sum \langle s, u_\alpha \rangle u_\alpha, u_\beta \rangle$$

$$= \langle s, u_\beta \rangle - \sum_\alpha \langle s, u_\alpha \rangle \langle u_\alpha, u_\beta \rangle$$

$$= \langle s, u_\beta \rangle - \langle s, u_\beta \rangle = 0$$

In the case of adaptive expansion, we do not use the orthonormal base but rather we use a new set of elementary function $\{h(t)\}$ to construct the mapping P. Assume that the subspace V can be accessed by a set of elementary function $\{h(t)\}$. Furthermore, assume that each $h(t)$ is normalized. Thus $V = \text{span}(H)$ where

$$H \subset \{h \in S, \|h(t)\| = 1\} \tag{A.40}$$

It is explicitly permitted that H be redundant, i.e., that it contains many more vectors than are needed to generate V. In other words, H may be linearly dependent. Now let $s(t)$ be any element of S. We wish to construct Ps, the point in V closest to s. The following algorithm serves this purpose.

Algorithm:

Given s, define $s_0 = s$. Select a number $\alpha \in (0,1)$. Proceed inductively. If s_n is known, select $h_n \in H$ so that

$$|\langle s_n, h_n \rangle| \geq \alpha \sup_{h \in H} |\langle s_n, h \rangle| \tag{A.41}$$

Define $v_n = \langle s_n, h_n \rangle h_n$ and $s_{n+1} = s_n - v_n$. v_n is the projection component of s_n, and s_{n+1} is the residuals component. Eq.(A.41) shows that we start with the largest of all projection components and then proceed to take smaller ones. Since

$$\|\langle s, h \rangle h\| = |\langle s, h \rangle| \|h\| = |\langle s, h \rangle| \tag{A.42}$$

we seek an $h_n \in H$ for which $|\langle s, h_n \rangle|$ is large. Ideally, we would select $h_n \in H$ so that $|\langle s, h_n \rangle|$ is a maximum. But such an h_n need not exist and a larger one is sufficient. This is why the constant α is present. One hopes that

$$Ps = \sum_{n=0}^{\infty} v_n = \sum_{n=0}^{\infty} \langle s_n, h_n \rangle h_n \tag{A.43}$$

We will show that is true. Notice that Eqs.(A.39) and (A.43) are quite similar in form, except in the former case, the expansion is done by an orthogonal base and is done using a redundant set of elementary functions.

Lemma 1 The vectors in the algorithm have these properties:

$$s_{n+1} \perp h_n, \; s_{n+1} \perp v_n \tag{A.44}$$

$$\|v_n\|^2 = |\langle s_n, h_n \rangle|^2 = \|s_n\|^2 - \|s_{n+1}\|^2 \tag{A.45}$$

$$s_n = \sum_{i=n}^{m-1} v_i + s_m \qquad \text{for } m \geq n \tag{A.46}$$

$$\|s\|^2 = \sum_{i=0}^{n} \|v_i\|^2 + \|s_{n+1}\|^2 \tag{A.47}$$

PROOF

To prove (A.44) , simply write

$$\langle s_{n+1}, h_n \rangle = \langle s_n - v_n, h_n \rangle = \langle s_n, h_n \rangle - \langle v_n, h_n \rangle$$
$$= \langle s_n, h_n \rangle - \langle s_n, h_n \rangle \langle h_n, h_n \rangle = 0$$

To prove (A.45), use (A.44) and the *Pythagorean law*, i.e.,

$$\|s_n\|^2 = \|s_{n+1} + v_n\|^2 = \|s_{n+1}\|^2 + \|v_n\|^2 \tag{A.48}$$

To prove (A.46), write

$$s_n = (s_n - s_{n+1}) + (s_{n+1} - s_{n+2}) + \ldots + (s_{m-1} - s_m) + s_m$$
$$= v_n + v_{n+1} + \ldots + v_{m-1} + s_m$$

To prove (A.47), use the property (A.45) as follows

$$\|s\|^2 = (\|s_0\|^2 - \|s_1\|^2) + (\|s_1\|^2 - \|s_2\|^2) + \ldots + (\|s_n\|^2 - \|s_{n+1}\|^2) + \|s_{n+1}\|^2$$
$$= \|v_0\|^2 + \|v_1\|^2 + \ldots + \|v_n\|^2 + \|s_{n+1}\|^2$$

Lemma 2

$$|\langle v_i, s_m \rangle| \le \alpha^{-1} \|v_i\| \|v_m\|$$

PROOF

$$\|v_i\| \|v_m\| = |\langle s_i, h_i \rangle| |\langle s_m, h_m \rangle| \ge |\langle s_i, h_i \rangle| \alpha |\langle s_m, h_i \rangle|$$
$$= \alpha |\langle \langle s_i, h_i \rangle h_i, s_m \rangle| = \alpha |\langle v_i, s_m \rangle|$$

Lemma 3   If $\{w_n\} \in l^2$ and $w_n \ge 0$ for all n, then for each n,

$$\inf_{k > n} w_{k+1} \sum_{j=0}^{k} w_j = 0 \tag{A.49}$$

PROOF

Fix n and $\varepsilon > 0$. Put $z_k = \sum_{i=0}^{k} w_i$. Select $m > n$ so that $\sum_{i > m} w_i^2 < \varepsilon/2$. Select $p > m$ so that $w_p z_m < \varepsilon/2$. Select

$$j \in \{m+1, m+2, ..., p\}$$

so that

$$w_j = \min\{w_{m+1}, w_{m+2}, ..., w_p\}$$

Then

$$\inf_{k > n} w_k z_k \le w_j z_j \le w_j z_p = w_j \left(z_m + \sum_{i=m+1}^{p} w_i \right)$$

$$= w_j z_m + \sum_{i=m+1}^{p} w_j w_i \le w_p z_m + \sum_{i=m+1}^{p} w_i^2 \le \varepsilon/2 + \varepsilon/2$$

$$= \varepsilon$$

Since ε was arbitrary, we conclude that $\inf_{k > n} w_k z_k = 0$.

Lemma 4   If $q > k$, then the points s_k and s_q produced in the algorithm satisfy

$$\|s_k - s_q\|^2 \le \|s_k\|^2 - \|s_q\|^2 + 2\alpha^{-1} \|v_q\| \sum_{i=k}^{q-1} \|v_i\| \tag{A.50}$$

PROOF

Since

$$\|s_k\|^2 = \|(s_k - s_q) + s_q\|^2 = \|s_k - s_q\|^2 + 2\langle s_k - s_q, s_q \rangle + \|s_q\|^2 \qquad \text{(A.51)}$$

we have, with the help of Eq.(A.46) and Lemma 2,

$$\|s_k - s_q\|^2 - \|s_k\|^2 + \|s_q\|^2 = -2\langle s_k - s_q, s_q \rangle = -2\sum_{i=k}^{q-1} \langle v_i, s_q \rangle$$

$$\leq 2\sum_{i=k}^{q-1} |\langle v_i, s_q \rangle| \leq 2\alpha^{-1} \|v_q\| \sum_{i=k}^{q-1} \|v_i\|$$

Lemma 5 The sequence $\{s_n\}$ produced by the algorithm is a *Cauchy* sequence. In other words, $\|s_n\|$ converges as n approaches infinity.

PROOF

Let $\varepsilon > 0$. By (A.45), it is apparent that the sequence $\|s_n\|$ is monotonically decreasing to a limit, $L \geq 0$. Select N so that

$$\|s_n\| < L^2 + \varepsilon^2$$

for all $n > k$. Let $m > n > N$. By (A.47),

$$\sum_{i=0}^{k} \|v_i\|^2 < \infty$$

Hence, Lemma 3 can be applied to obtain an index $q > m$ such that.

$$\|v_q\| \sum_{i=0}^{q} \|v_i\| < \varepsilon^2$$

By Lemma 4, for each k in the range $N < k < q$, we will have

$$\|s_k - s_q\|^2 \leq \|s_k\|^2 - \|s_q\|^2 + \frac{2}{\alpha} \|v_q\| \sum_{i-k}^{q-1} \|v_i\| < (L^2 + \varepsilon^2) - L^2 + \frac{2}{\alpha}\varepsilon^2 = \varepsilon^2 \left(1 + \frac{2}{\alpha}\right)$$

This leads to

$$\|s_k - s_q\| < \varepsilon \sqrt{\left(1 + \frac{2}{\alpha}\right)} \qquad \text{(A.52)}$$

Therefore, we have

$$\|s_m - s_n\| \le \|s_m - s_q\| + \|s_q - s_n\| < 2\varepsilon \sqrt{\left(1 + \frac{2}{\alpha}\right)} \tag{A.53}$$

With the above lemmas, we have the following theorem.

T H E O R E M

The series $\sum_{i=0}^{\infty} v_i$ produced by the algorithm converges to the orthogonal projection of s onto $V = \text{span}(H)$.

PROOF

By Lemma 5, the sequence $\{s_n\}$ is convergent and we are at liberty to define $y = \lim s_n$. By letting $n = 0$ and $m \to \infty$ in Eq.(A.46), we obtain

$$s = \sum_{i=0}^{\infty} v_i + y \tag{A.54}$$

By (A.46) and (A.53), the series $\sum_{i=0}^{\infty} v_i$ converges. Since $v_i \in V$ and V is closed, $\sum_{i=0}^{\infty} v_i \in V$. It remains to prove that $y \perp V$. By, $<s_n, h_n> \to 0$. For any $h \in H$, we have, by the definition of h_n,

$$|\langle s_n, h_n \rangle| \ge \alpha |\langle s_n, h \rangle| \tag{A.55}$$

Let $n \to 0$ in this inequality to get $<y, h> = 0$ for all $h \in H$. Hence $y \perp V$.

Joint Time-Frequency Analyzer

*I*n order to get a better understanding about the JTFA algorithms introduced, we include a demo disk of joint time-frequency analyzer introduced in the book. The joint time-frequency analyzer is a software-based instrument, developed by National Instruments Corporation, which allows users to analyze signals in time and frequency domains simultaneously. The analyzing signal could either be in the data file or acquired from outside via data acquisition boards. Except for the data-acquire portions that require extra hardware, this software basically preserves all other features of the original joint time-frequency analyzer. With this demo software, the reader can use different JTFA algorithms to test 11 built-in time-varying signals, such as the sum of multiple Gaussian functions, linear and non-linear chirp signals, biomedical samples, etc.

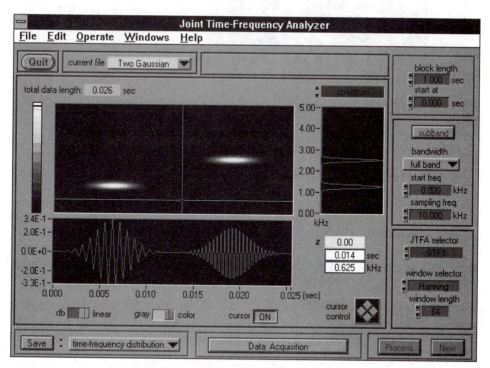

Fig. J–1 Front Panel of Joint Time-Frequency Analyzer

Fig. J–1 illustrates the front panel of the joint time-frequency analyzer. The operations of the Joint Time-Frequency Analyzer are rather straightforward.

The reader can press Ctrl+H to ask on-line help whenever he/she needs. The demo provides the following JTFA algorithms introduced in previous chapters:

- Adaptive spectrogram;
- Choi-Williams distribution;
- Cone-Shape distribution;
- Gabor spectrogram (or time-frequency distribution series);
- STFT spectrogram;
- Wigner-Ville distribution

In addition to these built-in algorithms, the complete joint time-frequency analyzer also allows users to design their own kernel for Cohen's class. But such capability is not available in this demo version.

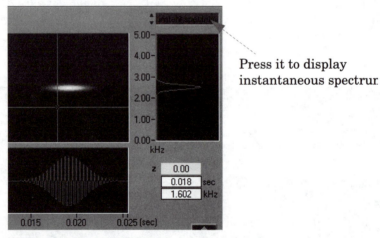

Press it to display instantaneous spectrum

Fig. J–2 The right plot displays the power spectrum at $t = 0.018$ sec. The time instant is controlled by the cursor.

Press "Save" to save   data selector
results into spreadsheet

Fig. J–3 Save the results into the spreadsheets.

As shown in Fig. J–1, the front panel mainly consists of three plots. The bottom one displays the time waveform. The right plot is the conventional power spectrum. By pressing the button, as shown in Fig. J–2, it will switch to the instantaneous spectrum, the profile of a spectrum at the particular time instant. The time is controlled by the cursor. The large plot in the top middle is the joint time-frequency representation, which roughly sketches the signal's time-frequency energy distribution. The reader could choose either linear or dB display. By pressing "Save," as shown in Fig. J–3, the reader could selectively save time waveform, spectrum, or joint time-frequency representation to a spreadsheet.

In what follows, we shall briefly compare the performance of different algorithms for different signals.

Sum of Two Gaussian Functions

The first example is the sum of two Gaussian functions, which is set as default. It is easy to see that the resolution of the STFT spectrogram is subject to the selection of the window function; the short window gives better time resolution and poor frequency resolution, or vice versa. It is impossible to accommodate both time and frequency resolution by the STFT spectrogram.

The Wigner-Ville distribution has much better time-frequency resolution. But it suffers from cross-term interference, in particular, when the WVD is computed from the real signal directly rather than using the corresponding analytical function.

For this example, we could find that the adaptive spectrogram and Gabor spectrogram have the best performance. When applying the Gabor spectrogram, the reader could start with zeroth order and then increase the order to one, two, three, etc. As the order increases, the resolution gets better. However, if the order is too large, say ten, then the cross-term starts to appear. A good compromise between the resolution and cross-term interference is usually of order two to four.

The above observations also apply for four Gaussian functions and the frequency hopper signal.

Linear Chirp with Gaussian Envelope

In this example, we would like to investigate the strengths and weaknesses of the adaptive spectrogram. Although the adaptive spectrogram yields a superior performance for the sum of multiple Gaussian functions and the frequency hopper signal, its result of the linear chirp signal is not so pleasant. This is because the elementary functions used in the Gaussian-based adaptive spectrogram are complex sine-function-modulated Gaussians, which are not efficient to match the rapid frequency changes, such as the chirp-type of signals. The Gaussian-based adaptive spectrogram is good at those signals whose frequency contents do not change very fast.

In this example, both Wigner-Ville distribution and Gabor spectrogram produce very good results.

Rectangular Pulse

By the example, we shall show the trade-off of using the analytical functions for the WVD. First, let's turn off the "Analytical" button to compute the Wigner-Ville distribution directly from the real sample. For this particular sample, the result obtained by the WVD is very good. It not only satisfies all useful properties, such as time marginal conditions, but also does not have cross-term interference. Next, press the "Analytical" button and compute the WVD with respect to the analytical function. As shown in Fig. J–4, there is severe distortion in the low-frequency band. Obviously, the analytical function's WVD does not satisfy the time marginal conditions. In most cases, such as multiple Gaussian functions and the frequency hopper signal, using the analytical functions, we could substantially reduce the cross-term interference of the WVD. On the other hand, as shown in Fig. J–4, the analytical function's WVD alters the original WVD, especially, in low frequency band.

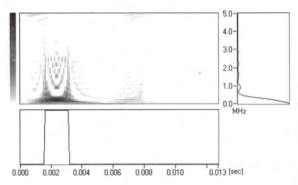

Fig. J–4 WVD of using analytical function (there are sever distortion in the low frequency band. Apparently, the analytical function's WVD do not satisfy the time marginal condition)

The reader could try other algorithms. To catch the abrupt jump, we have to use a short-window STFT spectrogram. If looking at the instantaneous spectrum, we will see that the resulting frequency resolution is very bad. However, if applying the fourth-order Gabor spectrogram with the wideband elementary function, we will see both good time and frequency resolutions.

Speech Signal

The speech signal usually is dominated by huge low-frequency components. To balance the high- and low-frequency portions, we often first apply the pre-emphasis filter, as shown in Fig. J–5. The range of the pre-emphasis parameter is 0 ~ 1. The larger the parameter is, the more the lower frequency portions are suppressed.

Because of the cross-term interference, neither the Wigner-Ville distribution nor the Choi-Williams distribution are suitable for speech analysis. The preferable algorithms for speech analysis are STFT spectrogram, Gabor spectrogram, and cone-shape distribution.

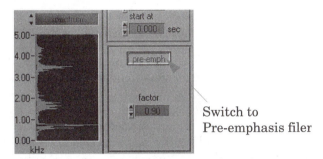

Switch to
Pre-emphasis filer

Fig. J–5 Apply the pre-emphasis filter to balance the low and high frequencies.

Biomedical Signal

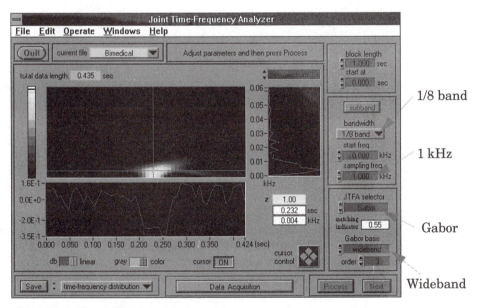

1/8 band

1 kHz

Gabor

Wideband

Fig. J–6 Subband filtering

In this example, the sampling rate is 1 kHz. But the majority of signals are below 50 Hz. To zoom in on a signal in the frequency domain, we could first apply a subband filter, as shown in Fig. J–6. In this case, we chose to display an eighth of the band. For most biomedical signals, neither STFT nor Gaussian-based adaptive spectrogram are good candidates. A good choice for a biomedical signal usually is the Gabor spectrogram or the Choi-Williams distribution. Due to the cross-term interference, the Wigner-Ville distribution is seldom used for biomedical signal analysis.

Summary

In addition to the examples discussed above, the reader can also test other samples provided by the demo with different parameters and algorithms. Every algorithm introduced in this book has its own strength and weakness. The selection of the algorithm to use is very much application-dependent. In general, however, the STFT and Gabor spectrogram (or time-frequency distribution series) are relatively robust and computationally efficient. In particular, the lower order Gabor spectrogram not only has a satisfactory resolution, but also substantially reduces the cross-term interference. Although it does not completely satisfy some useful properties, the difference often could be neglected in most applications.

Bibliography

The purpose of this bibliography is to provide the reader with sources for additional and more advanced treatments of topics in joint time-frequency analysis and its applications presented in Part 3. References are selected mainly on the basis of their relevance to the topics discussed in this book. This is by no means meant to be an exhaustive list but rather it is intended to indicate directions for further study.

[1] F. Auger and P. Flandrin, "Improving the readability of time-frequency and time-scale representations by the reassignment method," *IEEE Trans. Signal Processing*, vol. 43, no. 5, pp. 1068-1089, May 1995.

[2] D. A Ausherman, A. Kozma, J. L. Waker, H. M. Jones, and E. C. Poggio, "Developments in radar imaging," *IEEE Trans. Aerospace and Electronic Systems*, vol. 20, no. 4, pp. 363-400, 1984.

[3] L. Auslanders, I. C. Gertner, and R. Tolimieri, "The discrete zak transform application to time-frequency analysis and synthesis of nonstationary signals," *IEEE Trans. Signal Processing*, vol. 39, no. 4, pp. 825-835, April 1991.

[4] R. Balart, "Matrix reformulation of the Gabor transform," *Optical Engineering*, vol. 31, no. 6, pp. 1235-1242, June 1992.

[5] R. G. Baraniuk and D. L. Jones, "A signal-dependent time-frequency representation: Fast algorithm for optimal kernel design," *IEEE Trans. Signal Processing*, vol. 42, pp. 134-146, April 1993.

[6] R. G. Baraniuk and D. L. Jones, "A radially Gaussian, signal-dependent time-frequency representation: Optimal kernel design," *Signal Processing*, vol. 32, pp. 263-284, June 1993.

[7] R. G. Baraniuk and D. L. Jones, "A signal-dependent time-frequency representation," *IEEE Trans. Signal Processing*, vol. 41, pp. 1589-1602, January 1994.

[8] B. Bastiaans, "Gabor's expansion of a signal into Gaussian elementary signals," *Proc. IEEE*, vol. 68, pp. 538-539, April 1980.

[9] B. Bastiaans, "A sampling theorem for the complex spectrogram, and Gabor's expansion of a signal in Gaussian elementary signals," *Optical Engineering*, vol. 20, no. 4, pp. 594-598, July/August 1981.

[10] B. Bastiaans, "On the sliding-window representation in digital signal processing," *IEEE Trans. Acoustics, Speech, Signal Processing*, vol. ASSP-33, no. 4, pp. 868-873, August 1985.

[11] R. H. Bayley and P. M. Berry, "The electrical field produced by the eccentric current dipole in the nonhomogeneous conductor," *Am. Heart J.*, vol. 63, pp. 808-820, 1962.

[12] F. Black and M. Scholes, "The pricing of options and corporate liabilities," *Journal of Political Economy*, vol. 81, pp. 637-659, 1973.

[13] B. Boashash and P. J. Black, "An efficient real-time implementation of the Wigner-Ville distributions," *IEEE Trans. Acoustics, Speech, Signal Processing*, vol. 35, no. 11, pp.1611-1618, November 1987.

[14] B. Boashash, "Note on the use of the Wigner distribution for time-frequency signal

analysis," *IEEE Trans. Acoustics, Speech, Signal Processing*, vol. 36, no. 9, pp. 1518-1521, September 1988.

[15] G. F. Boudreaux-Bartels and T. W. Parks, "Time-varying filtering and signal estimation using Wigner distribution synthesis techniques," *IEEE Trans. Acoustics, Speech, Signal Processing*, vol. 34, pp. 442-451, 1986.

[16] A. C. Bovik, P. Maragos, and T. F. Quatieri, "AM-FM energy detection and separation in noise using multiband energy operators," *IEEE Trans. Signal Processing*, vol. 41, no. 12, pp. 3245-3265, December 1993.

[17] C. C. Chen and H. C. Andrews, "Target-motion-induced radar imaging," *IEEE Trans. Aerospace and Electronic Systems*, vol. 6, no. 1, pp. 2-14, 1980.

[18] P. Chen, "Empirical and theoretical evidence of monetary chaos," *System Dynamics Review*, vol. 4, pp. 81-108, 1988.

[19] P. Chen, "Study of chaotic dynamical systems via time-frequency analysis," *Proc. IEEE-SP Int. Symp. Time-Frequency and Time-Scale Analysis,* Philadelphia, October 25-28, pp. 357-360, 1994.

[20] P. Chen, "Deterministic cycles in evolving economy: time-frequency analysis of business cycles," in A. Aoki, K. Shiraiwa, and Y. Takahashi eds., *Dynamical Systems and Chaos, Vol. I. Mathematics, Engineering and Economics,* pp. 363-372. Singapore: World Scientific, 1995.

[21] V. Chen, "Radar ambiguity function, time-varying matched filter, and optimal wavelet correlator," *Optical Engineering*, vol. 33, no. 7, pp. 2212-2217, 1994.

[22] V. Chen, "Reconstruction of inverse synthetic aperture radar image using adaptive time-frequency wavelet transform," (invited paper), *SPIE Proc. Wavelet Applications*, vol. 2491, pp. 373-386, 1995.

[23] H. Choi and W. J. Williams, "Improved time-frequency representation of multicomponent signals using exponential kernels," *IEEE Trans. Acoustics, Speech, Signal Processing,* vol .37, no. 6, pp. 862-871, June 1989.

[24] C. K. Chui, *An Introduction to Wavelets*, New York: Academic Press, 1992.

[25] T. A. C. M. Claasen and W. F. G. Mecklenbräuker, "The Wigner Distribution–A tool for time-frequency signal analysis," *Phillips J. Res.*, vol. 35, pp. 217-250, pp. 276-300, pp. 1067-1072, 1980.

[26] L. Cohen, "Generalized phase-space distribution functions," *J. Math. Phys.*, vol. 7, pp. 781-806, 1966.

[27] L. Cohen and T. Posch, "Positive time-frequency distribution functions," *IEEE Trans. Acoustics, Speech, Signal Processing*, vol. 33, pp. 31-38, June 1985.

[28] L. Cohen, "On a fundamental property of the wigner distribution," *IEEE Trans. Acoustics, Speech, Signal Processing,* vol. 35, pp. 559-561, 1987.

[29] L. Cohen, "Wigner distribution for finite duration or band limited signals and limited cases," *IEEE Trans. Acoustics, Speech, Signal Processing*, vol. 35, pp. 796-806, 1987.

[30] L. Cohen, "Time-frequency distribution–A review," *Proc. IEEE*, vol. 77, no. 7, pp. 941-981, July 1989.

[31] L. Cohen and C. Lee, "Instantaneous frequency," in *Time-Frequency Signal Analysis*, edited by B. Boashash, pp. 98-117, Longman Cheshire, Wiley Halsted Press, 1992.

[32] L. Cohen, *Time-Frequency Analysis*, Englewood Cliffs, NJ: Prentice Hall,1995.

[33] B. N. Cuffin, "A method for localizing EEG sources in realistic head models," *IEEE Trans. Biomed. Engr.*, vol. 42, pp. 68-71, 1995.

[34] I. Daubechies, "Orthonormal bases of compactly supported wavelets," *Comm. Pure Appl. Math.*, vol. 4, pp. 909-996, November 1988.

[35] I. Daubechies, "The wavelet transform, time-frequency localization and signal analysis," *IEEE Trans. Information Theory*, pp. 961-1005, September 1990.

[36] I. Daubechies, *Ten Lectures on Wavelets*, Philadelphia: *SIAM*, 1992.

[37] I. Daubechies, H. Landau, and Z. Landau, "Gabor time-frequency lattices and the Wexler-Raz identity," submitted for publication.

[38] J. G. Daugman, "Complete discrete 2-D Gabor transforms by neural networks for image analysis and compression," *IEEE Trans. on Acoustics, Speech, Signal Processing*, vol. 36, no. 7, pp. 1169-1179, 1988.

[39] S. Deutsch and A. Deutsch, *Understanding the Nervous System, An Engineering Perspective*, Piscataway, NJ: IEEE Press, 1993.

[40] D. L. Donoho, "De-noise by soft thresholding," *IEEE Trans. Information Theory*, no. 41, pp. 613-627, May 1995.

[41] P. D. Einziger, "Gabor expansion of an aperture field in exponential elementary beams," *IEE Electron. Lett.*, vol. 24, pp. 665-666, 1988.

[42] G. Eichmann and B. Z. Dong, "Two-dimensional optical filtering of 1-D signals," *Appl. Opt.*, vol. 21, pp. 3152-3156, 1982.

[43] S. Farkash and S. Raz, "Linear systems in Gabor time-frequency space," *IEEE Trans. Signal Processing*, vol. 42, no. 3, pp. 611-617, March 1994.

[44] D. H. Fender, "Source localization of brain electrical activity," in *Handbook of Electroencephalography and Clinical Neurophysiology*, vol. 1, A. S. Gevins and A. Remond, eds., New York: Elsevier, pp. 355-403, 1987.

[45] P. Flandrin, "When is the Wigner-Ville spectrum non-negative?" in *Signal Processing III, Theories and Applications*, vol. 1, pp. 239-242, 1986.

[46] P. Flandrin, "On detection-estimation procedures in the time-frequency plane," *Proc. IEEE ICASSP-86*, vol. 4, pp. 2331-2334, 1986.

[47] P. Flandrin and F. Hlawatsch, "Signal representation geometry and catastrophes in the time-frequency plane," in *Mathematics in Signal Processing,* edited by T. S. Durrani, et al., Oxford: Clarendon, pp. 3-14, 1987.

[48] P. Flandrin, "Maximum signal energy concentration in a time-frequency domain," *Proc. IEEE ICASSP-88*, vol. 4, pp. 2176-2179, 1988.

[49] P. Flandrin, "A time-frequency formulation of optimal detection," *IEEE Trans. Acoustics, Speech, Signal Processing*, vol. 36, no. 9, pp. 1377-1384, September 1988.

[50] P. Flandrin and O. Rioul, "Affine smoothing of the Wigner-Ville distribution," in *Proc. IEEE ICASSP-90*, pp. 2455-2458, Albuquerque, NM, 1990.

[51] M. J. Freeman and M. E. Dunham, "Trans-ionospheric signal detection by time-scale representation," *UK Symposium on Applications of Time-Frequency and Time-Scale Methods, Proc. IEEE SP,* pp. 152-158.

[52] B. Friedlander and B. Porat, "Detection of transient signal by the Gabor representation," *IEEE Trans. Acoustics, Speech, Signal Processing*, vol. 37, no. 2, pp. 169-180, February 1989.

[53] D. Gabor, "Theory of communication," *J. IEE (London)*, vol. 93, no. III, pp. 429-457, November 1946.

[54] T. Genossar and M. Porat, "Can one evaluate the Gabor expansion using Gabor's iterative algorithm?" *IEEE Trans. Signal Processing*, vol. 40, no. 8, pp. 1852-1861, August 1992.

[55] A. Gersho, "Characterization of time-varying linear systems," *Proc. IEEE*, pp. 238, 1963.

[56] A. S. Givens, "Analysis of the electromagnetic signals of the human brain: Milestones, obstacles and goals," *IEEE Trans. Biomed. Engr.*, vol. 33, pp. 833-850, 1984.

[57] J. P. Gollub and H. L. Swinney, "Onset of turbulence in a rotating fluid," *Physics Review Letters*, vol. 35, pp. 929-930, 1975.

[58] G. H. Golub and C. F. Van Loan, *Matrix Computations*, Second Edition, Baltimore

and London: The Johns Hopkins University Press, 1989.

[59] P. Goupillaud, A. Grossmann, and J. Morlet, "Cycle-octave and related transformation in seismic signal analysis," *Geoexploration*, vol. 23, pp. 85-102, 1984.

[60] D. W. Griffin and J. S. Lim, "Signal estimation from modified short-time Fourier transform," *IEEE Trans. Acoustics, Speech, Signal Processing*, vol. 32, no. 2, pp. 236-243, April 1984.

[61] A. Grossman and J. Morlet, "Decomposition of hardy functions into square integrable wavelets of constant shape," *SIAM J. Math. Anal.*, vol. 15, pp. 723-736, 1984.

[62] M. Hamalainen, R. Hari, R. J. Ilmoniemi, J. Knuutila, and O.V. Lounasmaa, "Magnetoencephalography—Theory, instrumentation, and applications to noninvasive studies of the working human brain," *Reviews of Modern Physics*, vol. 65, pp. 413-497.

[63] M. S. Hamalainen and J. Sarvas, "Realistic conductivity geometry model of the human head for interpretation of neuromagnetic data," *IEEE Trans. Biomed. Engr.*, vol. 36, pp. 165-171, 1989.

[64] B. Harms, "Computing time-frequency distributions," *IEEE Trans. Signal Processing*, vol. 39, no. 3, pp. 727-729, March 1991.

[65] B. L. Hao, *Chaos II*, Singapore: World Scientific, 1990.

[66] R. M. Harper, R. J. Sclabassi, and T. Estrin, "Time series analysis and sleep research," *IEEE Trans. Autom. Contr.*, vol. AC-19,6, pp. 932-943, 1974.

[67] P. Hauri, *The Sleep Disorders*, 2nd Ed., Kalamazoo, MI: The Upjohn Company, 1982.

[68] C. Heil and D. Walnut, "Continuous and discrete wavelet transforms," *SIAM Rev.*, vol. 31, pp. 628-666, 1989.

[69] F. Hlawatsch, "Interference terms in the Wigner distribution," *Digital Signal Processing—84*, V. Cappellini and A. G. Constantinides (eds), Amsterdam: Elsevier Science Publishers, B. V. (North-Holland), pp. 363-367, 1984.

[70] F. Hlawatsch and W. Krattenthaler, "Phase matching algorithms for Wigner distribution signal synthesis," *IEEE Trans. Signal Processing*, vol. 39, no. 3, pp. 612-619, March 1991.

[71] F. Hlawatsch, "Duality and classification of bilinear time-frequency signal representations," *IEEE Trans. Signal Processing*, vol. 39, no. 7, pp. 1564-1574, July 1991.

[72] F. Hlawatsch and W. Krattenthaler, "Bilinear signal synthesis," *IEEE Trans. Signal Processing*, vol. 40, no. 2, pp. 352-363, February 1992.

[73] F. Hlawatsch and G.F. Boudreaux-Bartels, "Linear and quadratic time-frequency signal representations," *IEEE Signal Processing Magazine*, vol. 9, pp. 21-67, 1992.

[74] F. Hlawatsch, "The Wigner distribution of a linear signal space," *IEEE Trans. SP*, vol. 41, no. 3, pp. 1248-1258, March 1993, .

[75] F. Hlawatsch, A. H. Costa, and W. Krattenthaler, "Time-frequency signal synthesis with time-frequency extrapolation and don't-care regions," *IEEE Trans. Signal Processing*, vol. 42, no. 9, pp. 2513-2520, September 1994.

[76] F. Hlawatsch and P. Flandrin, "The interference structure of the Wigner distribution and related time-trequency signal representations," in *The Wigner Distribution— Theory and Applications in Signal Processing*, W. F. G. Mecklenbräuker, ed. Amsterdam: Elsevier, 1994.

[77] F. Hlawatsch and W. Kozek, "Time-frequency projection filters and time-frequency signal expansions," *IEEE Trans. Signal Processing*, vol. 42, no. 12, pp. 3321-3334, December 1994.

[78] F. Hlawatsch, T. G. Manickam, R.L. Urbanke, and W. Jones, "Smoothed pseudo-Wigner distribution, Choi-Williams distribution, and cone-kernel representation: Ambiguity-domain analysis and experimental comparison," *Signal Processing*, vol. 43, no. 2, pp. 149-168, May 1995.

[79] F. Hlawatsch, G. S. Edelson, and P. Podlucky, "Multipulse maximum-likelihood range/Doppler estimation and the ambiguity function of a linear signal space, *"IEEE Trans. Aerospace and Electronic Systems*, submitted.

[80] F. Hlawatsch and W. Krattenthaler, "Signal synthesis algorithms for bilinear time-frequency signal representations" in *The Wigner Distribution—Theory and Applications in Signal Processing*, ed. W. Mecklenbr, Ch. 3, pp. 135-209, Elsevier, to appear 1996.

[81] Hodrick and Prescott, "Post War US business cycles: an empirical investigation," Discussion paper No.451, Carnegie-Mellon University, 1981.

[82] R. A. Horn and C. R. Johnson, *Matrix Analysis*. Cambridge University Press, 1991.

[83] L. D. Jacobson and H. Wechsler, "The Wigner distribution and its usefulness for 2-D image processing," *Proc. 6th Int. Conf. on Pattern Recognition*, Munich, October 19-22, 1982.

[84] L. D. Jacobson and H. Wechsler, "The composite pseudo Wigner distribution (CSWD): A computable and versatile approximation to the Wigner distribution (WD)," *Proc. IEEE ICASSP-83*, pp. 254-256, 1983.

[85] L. D. Jacobson and H. Wechsler, "Joint spatial/spatial-frequency representation," *Signal Processing*, vol. 14, pp. 37-66, January 1988.

[86] A. J. E. M. Janssen, "Gabor representation of generalized functions," *J. Math. Anal. Appl.*, vol. 38, pp. 377-394, 1981.

[87] A. J. E. M. Janssen, "Positively properties of phase-plane disrtibution functions," *J. Math. Phys.*, vol. 25, pp. 2240-2252, July 1984.

[88] A. J. E. M. Janssen and T. A. C. M. Claassen, "On positively of time-frequency distributions," *IEEE Trans. Acoustics, Speech, Signal Processing*, vol. 33, no. 4, pp. 1029-1032, 1985.

[89] A. J. E. M. Janssen, "A note on positive time-frequency distributions," *IEEE Trans. Acoustics, Speech, Signal Processing*, vol. 35, no. 5, pp. 701-703, May 1987.

[90] A. J. E. M. Janssen, "On the locus and spread of pseudo-density functions in the time-frequency plane," *Phillips J. Res.*, vol. 37, no. 3, pp. 79-110, 1982.

[91] A. J. E. M. Janssen, "The Zak transform: A signal transform for sampled time-continuous signals," *Phillips J. Res.*, vol. 43, pp. 23-69, 1988.

[92] A. J. E. M. Janssen, "Positively of time-frequency distribution functions," *Signal Processing*, vol. 14, pp. 243-252, 1988.

[93] A. J. E. M. Janssen, "Optimality property of the Gaussian window spectrogram," *IEEE Trans. Signal Processing*, vol. 39, no. 1, pp .202-204, January 1991.

[94] A. J. E. M. Janssen, "Duality and biorthogonality for Weyl-Heisenberg frame," to appear in *J. Fourier Anal. Appl.*.

[95] A. J. E. M. Janssen, "On rationally oversampling Weyl-Heisenberg frame," submitted to *Signal Processing*.

[96] J. Jeong and W. Williams, "Alias-free generalized discrete-time time-frequency distributions," *IEEE Trans. Signal Processing*, vol. 40, no. 11, pp.2757-2765, November 1992.

[97] J. Jeong and W. Williams, "Kernel design for reduced interference distributions," *IEEE Trans. Signal Processing*, vol. 40, pp. 402-412, 1992.

[98] D. J. Jones and R. G. Baraniuk, "A simple scheme for adaptive time-frequency representations," *IEEE Trans. Signal Processing*, vol. 42, pp. 3530-3535, December 1994.

[99] D. J. Jones and R. G. Baraniuk, "An adaptive optimal-kernel time-frequency representation," *IEEE Trans. Signal Processing*, vol. 43, pp.2361-2371, October 1995.

[100] E. R. Kandel and J. H. Schwartz, *Principles of Neural Science*. New York: Elsevier/North-Holland, 1981.

[101] R. N. Kavanagh, T. M. Darcey, D. Lehmann, and D. H. Fender, "Evaluation of methods for three-dimensional localization of electrical sources in the human brain," *IEEE Trans. Biomed. Engr.*, vol. BME-24, pp. 421-429, 1978.

[102] N. Kawabata, "A nonstationary analysis of the electroencephalogram," *IEEE Trans. Biomed. Engr.*, vol. 20, pp. 444-452, 1973.

[103] S. M. Kay and S. L. Marple, "Spectrum analysis—a modern perspective," *Proc. IEEE*, vol. 69, pp. 1380, 1981.

[104] P. Kellaway and I. Petersen, *Quantitative Analytic Studies in Epilepsy*. New York: Raven Press, 1976.

[105] Hyeongdong Kim and Hao Ling, "Wavelet analysis of radar echo from finite-size targets," *IEEE Trans. Antennas Propagation*, vol. 41, no. 2, pp. 200-207, February 1993.

[106] P. J. Kootsookos, B. C. Lovell, and B. Boashash, "A unified approach to the STFT, TFD's, and instantaneous frequency," *IEEE Trans. Signal Processing*, vol. 40, no. 8, pp. 1971-1982, August 1992.

[107] W. Kozek and F. Hlawatsch, "A comparative study of linear and nonlinear time-frequency filters," *Proc. IEEE-SP Int. Symp. Time-Frequency Time-Scale Analysis*, Victoria, B.C. (Canada), pp. 163-166, October 1992.

[108] M. L. Kramer and D. L. Jones, "Improved time-frequency filtering using an STFT analysis-modification-synthesis method," in *Proc. Symp. Time-Frequency and Time-Scale Analysis*, Philadelphia, PA, pp. 264-267, 1994.

[109] W. Krattenthaler and F. Hlawatsch, "Improved signal synthesis from pseudo Wigner distribution," *IEEE Trans. Signal Processing*, vol. 39, no. 2, pp. 506-509, February 1991.

[110] W. Krattenthaler and F. Hlawatsch, "Time-frequency design and processing of signals via smoothed Wigner distribution," *IEEE Trans. Signal Processing*, vol. 41, no. 1, pp. 278-287, January 1993.

[111] S. Li, "The theory of frame multiresolution analysis and its applications," Ph.D dissertation, University of Maryland, May 1993.

[112] S. Li, "A general theory of discrete Gabor expansion," *Proc. of SPIE'94 Mathematical Imaging: Wavelet Applications*, San Diego, July 1994.

[113] S. Li, "General Frame decompositions, pseudo-duals and its application to Weyl-Heisenberg frames," to appear in *Numerical Functional Analysis and Optimization*.

[114] S. Li, "Non-separable 2-D discrete Gabor expansion for image processing," submitted to Special Issue of *Multidimensional Signal Processing*.

[115] S. Li and D. M. Healy Jr. "A parametric class of discrete Gabor expansions," *IEEE Trans. Signal Processing*, vol. 44, no. 2, pp. 201-211, February 1996.

[116] S. Li and S. Qian, "A complement to a derivation of discrete Gabor expansion," *IEEE Signal Processing Lett.*, vol. 2, no. 2, pp. 31-33, February 1995.

[117] W. Li, "Wigner distribution method equivalent to dechirp method for detecting a chirp signal," *IEEE Trans. Acoustics, Speech, Signal Processing*, vol. 35, no. 8, August 1988.

[118] Y. Lu and J. M. Morris, "On discrete Gabor transform," submitted for publication.

[119] S. Mallat, "A theory for multiresolution signal decomposition: The wavelet representation," *IEEE Trans. Pattern Anal., Machine Intell.*, vol. 11, pp. 674-693, July 1989.

[120] S. Mallat, "Multifrequency channel decompositions of images and wavelet models," *IEEE Trans. Acoustics, Speech, Signal Processing*, vol. 37, pp. 2091-2110, 1989.

[121] S. Mallat, "Multiresolution approximations and wavelet orthonormal bases of L2(R)," *Trans. of Amer. Math. Soc.*, vol. 315, pp. 69-87, September 1989.

[122] S. Mallat and Z. Zhang, "Matching pursuit with time-frequency dictionaries," *IEEE Trans. Signal Processing*, vol. 41, no. 12, pp. 3397-3415, December 1993.

[123] S. L. Maple, *Digital Spectral Analysis*. Englewood Cliffs, NJ: Prentice Hall, 1987.

[124] D. L. Mensa, *High Resolution Radar Imaging*. Dedham, MA: Artech House, 1981.

[125] Y. Meyer, "Ondelettes et functions splines," *Seminaire EDS*, Ecole Polytechnique, Paris, 1986.

[126] Y. Meyer, *Ondelettes et operateurs, I: Ondelettes, II: Operateurs de Calderon-Zygmund, III: Operateurs multilineaires*, Paris: Hermann, 1990.

[127] J. M. Morris and D. Wu, "Some results on joint time-frequency representation via Wigner-Ville distribution decomposition," *Proc. 1992 Conf. Inform. Sci. & Sys.*, vol. 1, pp. 6-10, 1992.

[128] J. M. Morris and D. Wu, "On alias-free formulation of discrete-time Cohen's class of distributions," submitted for publication.

[129] J. C. Mosher, P. S. Lewis, and R. M. Leahy, "Multiple dipole modeling and localization from spatio-temporal MEG data," *IEEE Trans. Biomed. Engr.*, vol. 39, pp. 541-552, 1992.

[130] J. C. de Munck, "The potential distribution in a layered anisotropic spheroidal volume conductor," *J. Appl. Phys.*, vol. 64, pp. 464-470, 1988.

[131] S. H. Nawab and T. F. Quatieri, "Short-time Fourier transform," in *Advanced Topics in Signal Processing*, edited by J.S. Lim and A.V. Oppenheim. Englewood Cliffs, NJ: Prentice Hall, 1988.

[132] B. Noble and J. W. Daniel, *Applied Linear Algebra*. 3rd Ed., Chapter 9, pp. 355-395. Englewood Cliffs, NJ: Prentice Hall, 1988.

[133] A. H. Nuttall, "Wigner distribution function: relation to short-term spectral estimation, smoothing, and performance in noise," Naval Underwater Systems Center, *Technical Report 8225*, 1988.

[134] A. H. Nuttall, "The Wigner distribution function with minimum spread," Naval Underwater Systems Center, *Technical Report 8317*, 1988.

[135] A. H. Nuttall, "Alias-free Wigner distribution function and complex ambiguity function for discrete signals," Naval Underwater Systems Center, *Technical Report 8533*, 1989.

[136] A. V. Oppenheim and R. W. Schafer, *Digital Signal Processing*, Englewood Cliffs, NJ: Prentice Hall, 1975.

[137] R. S. Orr, "The order of computation of finite discrete Gabor transforms," *IEEE Trans. Signal Processing*, vol. 41, no. 1, pp. 122-130, January 1993.

[138] R. S. Orr, "Derivation of Gabor transform relatios using bessel's equality," *Signal Processing*, vol. 30, pp. 257-262, 1993.

[139] C. H. Page, "Instantaneous power spectrum," *J. Appl. Phys.*, vol. 23, pp. 103-106, 1952.

[140] A. Papandreou and G. F. Boudreaux-Bartels, "A generalization of the Choi-Williams and the Butterworth time-frequency distributions," *IEEE Trans. Signal Processing*, vol. 41, pp. 463-472, 1993.

[141] A. Papoulis, *Signal Analysis*, New York: McGraw-Hill, 1977.

[142] S. C. Pei and T. Y. Wang, "The Wigner distribution of linear time-variant systems," *IEEE Trans. Acoustics, Speech, Signal Processing*, vol. 36, pp. 1681-1684, 1988.

[143] T. Posch, "Wavelet transform and time-frequency distributions," in *Proc. SPIE, Int. Soc. Optical Engineering*, vol. 1152, pp. 477-482, 1988.

[144] M. R. Protnoff, "Time-frequency representation of digital signal and systems based on short-time Fourier analysis," *IEEE Trans. Acoustics, Speech, Signal Processing*, vol. 28, no. 1, pp. 55-69, February, 1980.

[145] S. Qian and J. M. Morris, "A fast algorithm for real joint time-frequency transformation of time-varying signals," *Electronic Let.*, vol. 26, pp. 537-539, 1990.

[146] S. Qian, D. Chen, and K. Chen, "Signal approximation via data-adaptive normalized Gaussian functions and its applications for speech processing," *Proc. ICASSP-92*, San Francisco, CA, March 23-26, 1992, pp. 141-144.

[147] S. Qian and J. M. Morris, "Wigner distribution decomposition and cross-term deleted representation," *Signal Processing*, vol. 25, no. 2, pp. 125-144, May 1992.

[148] S. Qian, K. Chen and S. Li, "Optimal biorthogonal sequence for finite discrete-time Gabor expansion," *Signal Processing*, vol. 27, no. 2, pp. 177-185, May 1992.

[149] S. Qian and D. Chen, "Discrete Gabor transform," *IEEE Trans. Signal Processing*, vol. 41, no. 7, pp. 2429-2439, July 1993.

[150] S. Qian and D. Chen, "Optimal biorthogonal analysis window function for discrete Gabor transform," *IEEE Trans. Signal Processing*, vol. 42, no. 3, pp. 687-694, March 1994.

[151] S. Qian and D. Chen, "Signal representation using adaptive normalized Gaussian functions," *Signal Processing*, vol. 36, no. 1, pp. 1-11, March 1994.

[152] S. Qian and D. Chen, "Decomposition of the Wigner-Ville distribution and time-frequency distribution series" *IEEE Trans. Signal Processing*, vol. 42, no. 10, pp. 2836-2841, October 1994.

[153] S. Qian, M. E. Dunham, and M. J. Freeman, "Trans-ionospheric signal recognition by joint time-frequency representation," *Radio Science*, vol. 30, no. 6, pp. 1817-1829, November-December 1995.

[154] S. Qiu and H. G. Feichtinger, "Discrete Gabor structure and optimal representation," *IEEE Trans. Signal Processing*, vol. 43, no. 10, pp. 2258-2268, October 1995.

[155] S. Qiu, "Block-circulant Gabor-matrix structure and discrete Gabor transforms," *Optical Engineering*, vol. 34, no. 10, pp. 2872-2878, October 1995.

[156] W. Rihaczek, "Signal energy distribution in time and frequency," *IEEE Trans. Inst. Radio Engineers (IRC)*, vol. IT-14, pp. 369-374, 1968.

[157] O. Rioul, "Wigner-Ville representations of signals adapted to shifts and dilations," *Tech. Memo. 112277-880422-03-TM*, AT&T Bell Labs, 1988.

[158] O. Rioul and M. Vetterli, "Wavelets and signal processing," *IEEE Signal Processing Magazine*, pp. 14-39, October 1991.

[159] O. Rioul and P. Flandrin, "Time-scale energy distributions: A general class extending wavelet transforms," *IEEE Trans. SP*, vol. 40, pp. 1746-1757, 1992.

[160] O. Rioul and P. Duhamel, "Fast algorithms for discrete and continuous wavelet transform," *IEEE Trans. Inform. Theory*, vol. 38, pp. 569-586, 1992.

[161] B. E. Saleh and N. S. Subotic, "Time-variant filtering of signals in the mixed time-frequency domain," *IEEE Trans. Acoustics, Speech, Signal Processing*, vol. 33, no. 3, pp. 1479-1485, 1985.

[162] R. J. Sclabassi and R. M. Harper, "Laboratory computers in neurophysiology," *Proc. IEEE*, vol. 61, no. 11, pp. 1602-1614, 1973.

[163] R. J. Sclabassi, M. Sun, D. N. Krieger, P. Jasiukaitis, and M. S. Scher, "Time-frequency analysis of the EEG signal," in *Proc. of ISSP 90, Signal Processing, Theories, Implementations and Applications*, Gold Coast, Australia, pp. 935-942, 1990.

[164] R. J. Sclabassi, M. Sun, D. N. Krieger, P. Jasiukaitis, and M. S. Scher, "Time-frequency domain problems in the neurosciences," in *Time-Frequency Signal Analysis: Methods and Applications*, editor B. Boashash, Longman-Cheshire, Wiley Halsted Press, pp. 498-519, 1992.

[165] M. I. Skolnik, *Introduction to Radar Systems*, 2nd edition,New York: McGraw-Hill, 1980.

[166] M. Steriade, D. A. McCormick, and T. J. Sejnwski, "Thalamocortical oscillations in the sleeping and aroused brain," *Science*, vol. 262, pp. 679-685, 1993.

[167] G. Strang, "Wavelets and dilation equations: A brief introduction," SIAM Rev., vol. 31, pp. 614-627, December 1989.

[168] G. Strang and T. Q. Nguyen, *Wavelet and Filter Banks*. Wellesley, MA: Wellesley-Cambridge Press, 1995.

[169] M. Sun, C.C. Li, L.N. Sekhar, and R.J. Sclabassi, "Efficient computation of the discrete pseudo-Wigner distribution," *IEEE Trans. Acoustics, Speech, Signal Processing*, vol. 37, pp. 1735-1742, 1989.

[170] M. Sun and R. J. Sclabassi, "Discrete instantaneous frequency and its computation," *IEEE Trans. Signal Processing*, vol. 41, pp. 1867-1880, 1993.

[171] M. Sun, F-C. Tsui, and R. J. Sclabassi, "Partially reconstructible wavelet decomposition of evoked potentials for dipole dource localization," in *Proc. 15th Annual Int. Conf., IEEE Engr. in Medicine and Biology Soc.*, San Diego, pp. 332-333, 1993.

[172] M. Sun, F-C. Tsui, and R. J. Sclabassi, "Multiresolution EEG source localization using the wavelet transform," In *Proc. of the IEEE 19th Northeast Biomedical Engineering Conf.*, Newark, NJ, pp. 88-91, March 1993.

[173] S. M. Sussman, "Least squares synthesis of radar ambiguity functions," *Trans. Inst. Radio Engineers (IRE)*, vol. 8, pp. 246-254, 1962.

[174] H. H. Szu and J. A. Blodgett, "Wigner distribution and ambiguity function," in *Optics in Four Dimensions,* edited by L.M. Narducci, pp. 355-381. New York: American Institute of Physics, 1981.

[175] H. H. Szu, "Two dimensional optical processing of one-dimensional acoustic data," *Optical Engineering,* vol. 21, pp. 804-813, 1982.

[176] H. H. Szu, "Signal processing using bilinear and nonlinear time-frequency joint-distributions," in *The Physics of Phase Space*, edited by Y. S. Kim and W. W. Zachary, New York: Springer Verlag, pp. 179-199, 1987.

[177] R. Tolimieri and R. S. Orr, "Poisson summation, the ambiguity function and the theory of Weyl-Heisenberg frames," *IEEE Trans. Inst. Radio Engineers (IRC)*, 1995.

[178] L. C. Trintinalia and H. Ling, "Extraction of waveguide scattering features using joint time-frequency ISAR," *IEEE Microwave Guided Wave Lett.*, vol. 6, pp. 10-12, January 1996.

[179] P. P. Vaidyanathan, *Multirate Systems and Filter Banks*, Englewood Cliffs, NJ: Prentice Hall, 1993.

[180] M. Vetterli and J. Kovacevic, *Wavelets and Subband Coding*. Englewood Cliffs, NJ: Prentice Hall, 1995.

[181] J. Ville, "Thovrie et applications de la notion de signal analylique," (in French) *Câbles et Transmission*, vol. 2, pp. 61-74, 1948.

[182] H. T. G. Wang, M. L. Sanders, and A. Woo, "Radar cross section measurement data of the VFY 218 configuration," *Tech. Rept. NAWCWPNS TM-7621*, Naval Air Warfare Center, China Lake, CA, January 1994.

[183] J. P. C. de Weerd and J. I. Kap, "Spectro-temporal representations and time-varying spectra of evoked potentials: A methodological investigation," *Biol. Cybern.*, vol. 41, pp. 101-117, 1981.

[184] D. R. Wehner, *High-Resolution Radar*, Second Edition. Boston-London: Artech House, 1995.

[185] J. Wexler and S. Raz, "Discrete Gabor expansions," *Signal Processing*, vol. 21, no. 3, pp. 207-221, November 1990.

[186] Mladen Victor Wickerhauser, *Adapted Wavelet Analysis from Theory to Software*. Wellesley, MA: A. K. Peters, 1994.

[187] E. P. Wigner, "On the Quantum Correction for Thermodynamic Equilibrium," *Phys. Rev.*, vol. 40, pp. 749, 1932.

[188] E. P. Wigner, "Quantum-mechanical distribution functions revisited," *Perspectives in Quantum Theory*, W. Yourgrau and A. van der Merwe, eds., pp. 25-36, Cambridge, MA: MIT Press, 1971.

[189] W. Williams, H. P. Zaveri, and J. C. Sackellares, "Time-frequency analysis of electrophysiology signals in epilepsy," *IEEE Trans. Engr. Med. Biol.*, pp. 133-143, March/April 1995.

[190] P. M. Woodward, *Probability and Information Theory with Applications to Radar*. Elmsford, NY: Pergamon Press, 1953.

[191] D. Wu and J. M. Morris, "Time-frequency representations using a radial Butterworth kernel," *Proc. IEEE-SP Int. Symp. on Time-Frequency and Time-Scale Analysis*, pp. 60-63, Philadelphia, PA, October 25-28, 1994.

[192] XiangGen Xia, "Topics in wavelet transforms," Ph.D. dissertation, University of Southern California, 1992.

[193] XiangGen Xia, C.-C. Jay Kuo, and Z. Zhang, "Wavelet coefficient computation with optimal prefiltering," *IEEE Trans. Signal Processing*, vol. 42, pp. 2191-2197, 1994.

[194] XiangGen Xia and S. Qian, "Gabor expansion based time-variant filter," submitted for publication.

[195] J. Yao, "Complete Gabor transformation for signal representation," *IEEE Trans. Image Processing*, vol. 2, pp. 152-159, April 1993.

[196] B. Zhang and S. Sato, "A time-frequency distribution of Cohen's class with a compound kernel and its application to speech signal processing," *IEEE Trans. Signal Processing,* vol. 42, no. 4, pp. 54-64, January 1994.

[197] Y. Zhao, L. E. Atlas, and R. J. Marks, "The use of cone-shaped kernels for generalized time-frequency representations of nonstationary signals," *IEEE Trans. Acoustics, Speech, Signal Processing*, vol. 38, no. 7, pp. 1084-1091, July 1990.

[198] M. Zibulski and Y. Y. Zeevi, "Oversampling in the Gabor scheme," *IEEE Trans. Signal Processing*, vol. 41, no. 8, August 1993.

[199] Zibulski and Y. Y. Zeevi, "Frame analysis of the discrete Gabor-scheme," *IEEE Trans. Signal Processing*, vol. 42, no. 4, pp. 942-943, April 1994.

Index

A

adaptive Gabor representation (AGR), 190, 234
adaptive spectrogram, *See* spectrogram
admissibility condition, 80
affine correlation, 120
analysis function, 16, 56
analytical signal, 123, 284
augmented matrix, 266
auto-correlation function
    auto-correlation function, 22
    time-dependent auto-correlation function, 103, 132, 139

B

backscattering feature extraction, 229
bandpass, 80, 244, 258
basis, 17
bilinear, 101, 121, 139, 142
biorthogonal, 17, 54, 65

C

Choi-Williams distribution, *See* Cohen's class
Cohen's class
    Choi-Williams distribution, 145, 174, 282
    Cohen's class, 139, 173
    cone-shape distribution, 146, 178, 282
    signal-dependent time-frequency representation, 149
    symmetric ambiguity function (AF), 132, 173
cone-shape distribution, *See* Cohen's class
constant Q, 36, 79, 193
continuous-time Fourier transform, *See* Fourier transform
continuous-time wavelet transform, *See* wavelets
critical sampling, 53, 265
cross-range, 215
cross-term interference, 112, 136, 162, 187, 245, 283

W

LICENSE AGREEMENT AND LIMITED WARRANTY

READ THE FOLLOWING TERMS AND CONDITIONS CAREFULLY BEFORE OPENING THIS DISK PACKAGE. THIS LEGAL DOCUMENT IS AN AGREEMENT BETWEEN YOU AND PRENTICE-HALL, INC. (THE "COMPANY"). BY OPENING THIS SEALED DISK PACKAGE, YOU ARE AGREEING TO BE BOUND BY THESE TERMS AND CONDITIONS. IF YOU DO NOT AGREE WITH THESE TERMS AND CONDITIONS, DO NOT OPEN THE DISK PACKAGE. PROMPTLY RETURN THE UNOPENED DISK PACKAGE AND ALL ACCOMPANYING ITEMS TO THE PLACE YOU OBTAINED THEM FOR A FULL REFUND OF ANY SUMS YOU HAVE PAID.

1.   **GRANT OF LICENSE:** In consideration of your payment of the license fee, which is part of the price you paid for this product, and your agreement to abide by the terms and conditions of this Agreement, the Company grants to you a nonexclusive right to use and display the copy of the enclosed software program (hereinafter the "SOFTWARE") on a single computer (i.e., with a single CPU) at a single location so long as you comply with the terms of this Agreement. The Company reserves all rights not expressly granted to you under this Agreement.

2.   **OWNERSHIP OF SOFTWARE:** You own only the magnetic or physical media (the enclosed disks) on which the SOFTWARE is recorded or fixed, but the Company retains all the rights, title, and ownership to the SOFTWARE recorded on the original disk copy(ies) and all subsequent copies of the SOFTWARE, regardless of the form or media on which the original or other copies may exist. This license is not a sale of the original SOFTWARE or any copy to you.

3.   **COPY RESTRICTIONS:** This SOFTWARE and the accompanying printed materials and user manual (the "Documentation") are the subject of copyright. You may not copy the Documentation or the SOFTWARE, except that you may make a single copy of the SOFTWARE for backup or archival purposes only. You may be held legally responsible for any copying or copyright infringement which is caused or encouraged by your failure to abide by the terms of this restriction.

4.   **USE RESTRICTIONS:** You may not network the SOFTWARE or otherwise use it on more than one computer or computer terminal at the same time. You may physically transfer the SOFTWARE from one computer to another provided that the SOFTWARE is used on only one computer at a time. You may not distribute copies of the SOFTWARE or Documentation to others. You may not reverse engineer, disassemble, decompile, modify, adapt, translate, or create derivative works based on the SOFTWARE or the Documentation without the prior written consent of the Company.

5.   **TRANSFER RESTRICTIONS:** The enclosed SOFTWARE is licensed only to you and may not be transferred to any one else without the prior written consent of the Company. Any unauthorized transfer of the SOFTWARE shall result in the immediate termination of this Agreement.

6.   **TERMINATION:** This license is effective until terminated. This license will terminate automatically without notice from the Company and become null and void if you fail to comply with any provisions or limitations of this license. Upon termination, you shall destroy the Documentation and all copies of the SOFTWARE. All provisions of this Agreement as to warranties, limitation of liability, remedies or damages, and our ownership rights shall survive termination.

7.   **MISCELLANEOUS:** This Agreement shall be construed in accordance with the laws of the United States of America and the State of New York and shall benefit the Company, its affiliates, and assignees.

8.   **LIMITED WARRANTY AND DISCLAIMER OF WARRANTY:** The Company warrants that the SOFTWARE, when properly used in accordance with the Documentation, will operate in substantial conformity with the description of the SOFTWARE set forth in the Docu-

mentation. The Company does not warrant that the SOFTWARE will meet your requirements or that the operation of the SOFTWARE will be uninterrupted or error-free. The Company warrants that the media on which the SOFTWARE is delivered shall be free from defects in materials and workmanship under normal use for a period of thirty (30) days from the date of your purchase. Your only remedy and the Company's only obligation under these limited warranties is, at the Company's option, return of the warranted item for a refund of any amounts paid by you or replacement of the item. Any replacement of SOFTWARE or media under the warranties shall not extend the original warranty period. The limited warranty set forth above shall not apply to any SOFTWARE which the Company determines in good faith has been subject to misuse, neglect, improper installation, repair, alteration, or damage by you. EXCEPT FOR THE EXPRESSED WARRANTIES SET FORTH ABOVE, THE COMPANY DISCLAIMS ALL WARRANTIES, EXPRESS OR IMPLIED, INCLUDING WITHOUT LIMITATION, THE IMPLIED WARRANTIES OF MERCHANTABILITY AND FITNESS FOR A PARTICULAR PURPOSE. EXCEPT FOR THE EXPRESS WARRANTY SET FORTH ABOVE, THE COMPANY DOES NOT WARRANT, GUARANTEE, OR MAKE ANY REPRESENTATION REGARDING THE USE OR THE RESULTS OF THE USE OF THE SOFTWARE IN TERMS OF ITS CORRECTNESS, ACCURACY, RELIABILITY, CURRENTNESS, OR OTHERWISE.

IN NO EVENT, SHALL THE COMPANY OR ITS EMPLOYEES, AGENTS, SUPPLIERS, OR CONTRACTORS BE LIABLE FOR ANY INCIDENTAL, INDIRECT, SPECIAL, OR CONSEQUENTIAL DAMAGES ARISING OUT OF OR IN CONNECTION WITH THE LICENSE GRANTED UNDER THIS AGREEMENT, OR FOR LOSS OF USE, LOSS OF DATA, LOSS OF INCOME OR PROFIT, OR OTHER LOSSES, SUSTAINED AS A RESULT OF INJURY TO ANY PERSON, OR LOSS OF OR DAMAGE TO PROPERTY, OR CLAIMS OF THIRD PARTIES, EVEN IF THE COMPANY OR AN AUTHORIZED REPRESENTATIVE OF THE COMPANY HAS BEEN ADVISED OF THE POSSIBILITY OF SUCH DAMAGES. IN NO EVENT SHALL LIABILITY OF THE COMPANY FOR DAMAGES WITH RESPECT TO THE SOFTWARE EXCEED THE AMOUNTS ACTUALLY PAID BY YOU, IF ANY, FOR THE SOFTWARE.

SOME JURISDICTIONS DO NOT ALLOW THE LIMITATION OF IMPLIED WARRANTIES OR LIABILITY FOR INCIDENTAL, INDIRECT, SPECIAL, OR CONSEQUENTIAL DAMAGES, SO THE ABOVE LIMITATIONS MAY NOT ALWAYS APPLY. THE WARRANTIES IN THIS AGREEMENT GIVE YOU SPECIFIC LEGAL RIGHTS AND YOU MAY ALSO HAVE OTHER RIGHTS WHICH VARY IN ACCORDANCE WITH LOCAL LAW.

ACKNOWLEDGMENT

YOU ACKNOWLEDGE THAT YOU HAVE READ THIS AGREEMENT, UNDERSTAND IT, AND AGREE TO BE BOUND BY ITS TERMS AND CONDITIONS. YOU ALSO AGREE THAT THIS AGREEMENT IS THE COMPLETE AND EXCLUSIVE STATEMENT OF THE AGREEMENT BETWEEN YOU AND THE COMPANY AND SUPERSEDES ALL PROPOSALS OR PRIOR AGREEMENTS, ORAL, OR WRITTEN, AND ANY OTHER COMMUNICATIONS BETWEEN YOU AND THE COMPANY OR ANY REPRESENTATIVE OF THE COMPANY RELATING TO THE SUBJECT MATTER OF THIS AGREEMENT.

Should you have any questions concerning this Agreement or if you wish to contact the Company for any reason, please contact in writing at the address below or call the at the telephone number provided.

PTR Customer Service
Prentice Hall PTR
One Lake Street
Upper Saddle River, New Jersey 07458

Telephone: 201-236-7105